MOLECULAR
PHOTODISSOCIATION
DYNAMICS

Advances in
Gas-Phase Photochemistry and Kinetics
Edited by M. N. R. Ashfold and J. E. Baggott

Each title in this series aims to provide critical reviews on a specific area of current interest at the interface of molecular spectroscopy, photochemistry, and chemical kinetics.

Molecular Photodissociation Dynamics

How to obtain future titles on publication

'Molecular Photodissociation Dynamics' is the first title in a new series 'Advances in Gas-Phase Photochemistry and Kinetics'. The next title in this series will deal with gas-phase bimolecular association reactions. Future titles may be obtained as soon as they are published by sending a standing order to:

The Royal Society of Chemistry
Distribution Centre
Blackhorse Road
Letchworth
Herts. SG6 1HN

Telephone: Letchworth (0462) 672555
Telex: 825372

Advances in
Gas-Phase Photochemistry and Kinetics

Molecular Photodissociation Dynamics

M. N. R. Ashfold
School of Chemistry
University of Bristol

J. E. Baggott
Department of Chemistry
University of Reading

ROYAL
SOCIETY OF
CHEMISTRY

o 288- 7939

CHEMISTRY

British Library Cataloguing in Publication Data

Molecular photodissociation dynamics.—
  (Advances in gas-phase photochemistry and
  kinetics).
  1. Photoionization   2. Dissociation
  3. Molecules
  I. Ashfold, M. N. R.   II. Baggott, J. E.
  III. Royal Society of Chemistry   IV. Series
  539'.6   QD561

ISBN 0-85186-373-6

Front Cover Illustration
Schematic representation of the fragmentation of a water
molecule to yield an H atom and a highly rotationally
excited OH radical. Symmetry conservation dictates that
there will be selective formation of one (the upper) Λ-doublet
state of the OH product. (After Andresen and Schinke, Chapter 3.)

Published by The Royal Society of Chemistry
Burlington House, Piccadilly, London W1V 0BN

Typeset by Bath Typesetting Ltd., Bath,
and printed by J. W. Arrowsmith Ltd., Bristol, England

Foreword

The ready availability of tunable laser sources, operational over a substantial portion of the electromagnetic spectrum, has heralded nothing short of a revolution in the study of photodissociation. Consequent upon any revolution of this sort comes a proliferation of notable research papers in the scientific literature. Hence the need for the present volume, in which a number of internationally acknowledged experts in the field of gas-phase molecular photodissociation dynamics have been invited to provide overviews of the 'state of the art' in a few selected aspects of their subject.

The earliest studies of photodissociation were concerned with identifying the primary products. Through the ensuing decades this particular question has remained central to any thorough understanding of photofragmentation processes and, with present-day technology, we are now in a position to provide extremely detailed answers. The opening chapters of this book (by Bersohn and by Wodtke and Lee) survey many of the ways in which photofragment translational spectroscopy combined with mass spectrometric detection methods has been used to provide information concerning the primary products and, in the event of there being more than one dissociation channel, the relative product yields. Dynamical aspects of the dissociation processes occurring in a wide variety of molecules, ranging from diatomics to large polyatomics, have been explored using these methods. Though blessed with almost universal applicability and high sensitivity, these detection techniques currently lack the resolution necessary for the study of these processes at the level of full quantum-state definition. However, there exist a variety of laser-based detection schemes (most prominent among these being the technique of laser-induced fluorescence) capable of resolving this level of detail in (suitably chosen) small molecular fragments. Illustrative examples are described in Chapters 3–6.

Clearly it is sensible to strive hardest for full quantum-state definition in those systems for which theory is most ready to be tested or stretched. The water molecule has long been favoured as a 'model' system for those interested in polyatomic photofragmentation processes. Its virtues are further demonstrated in Chapter 3 (by Andresen and Schinke), which surveys the accord between experiment and *ab initio* theory relating to the direct dissociation of H_2O from its first excited $(\tilde{A}^1B_1)$ electronic state and

v

culminates in a description of the first fully quantum-state-selected photo-fragmentation.

The water molecule also features prominently in the first part of Chapter 4 (by Docker, Hodgson, and Simons), where, along with studies of NH_3, it is used to illustrate the importance of the initial parent quantum-state selection on the resulting photofragmentation dynamics. The focus switches in the remainder of the chapter to the kinds of dynamical information that can be obtained from careful study of the product excitation spectra. Doppler lineshape measurements can, in favourable cases, provide us with information on translational-energy release. Polarization studies yield information about product alignment (the correlation between the product rotational angular momentum vector, J, and the transition dipole, μ). Sometimes, as in the featured example of H_2O_2 photodissociation yielding OH fragments in their ground electronic states, it is even possible to identify effects on the individual lineshapes due to the mutual correlation between v, the product recoil velocity, and J: such measurements provide us with uniquely detailed answers to our questions concerning the forces acting as the recoiling fragments separate.

The relationship between photochemistry and spectroscopy is a recurring theme throughout this book. Nowhere is it more evident than in Chapter 5 (by Reisler, Noble, and Wittig), which compares and contrasts the fragmentation dynamics of two families of NO-containing parent molecules following excitation to their respective first excited singlet states. HONO and the alkyl nitrites dissociate rapidly, on the time-scale of a few vibrational periods, yielding fragments characterized by markedly non-statistical energy partitioning. In contrast, the fragmentation of HNO and the aliphatic nitroso compounds occurs slowly, only after radiationless transfer to the lower-lying triplet state and/or to high vibrational levels of the ground state. This allows ample time for complete energy randomization within the dissociating molecule with the consequence that (in cases where there is no significant exit channel barrier) the forms of the eventual product-state population distributions are well described using statistically based theories.

Chapter 6 (by Crim) highlights another means of studying molecular photodissociation processes that occur on a ground-state potential-energy surface. In this case energy is deposited directly in high vibrational levels by excitation of bond-stretching overtones, allowing a high degree of specificity to be introduced in the excited molecule. Because of the nature of the overtone vibrations of the larger polyatomics, the excitation is not strictly quantum-state selective. However, excitation of smaller polyatomics (*e.g.* H_2O_2) cooled in a supersonic expansion offers the prospect of obtaining much detailed information on the resulting dissociation processes.

We close in Chapter 7 (by Powis) with an overview of multi-photon ionization and its uses in providing well characterized sources of ions for subsequent spectroscopic or dynamical studies. The bulk of this chapter is concerned with discussion of the mechanisms of ion fragmentation following photon absorption and with the ionization of resonant intermediates formed in the multi-photon process. Multi-photon ionization detection schemes will surely continue to find more applications, and the prospects for the technique are discussed with reference to studies of unimolecular dissociation, photofragment spectroscopy, and ion–molecule reactions.

Evolving trends, most notably the rapid growth of the scientific literature and the increased availability of computer-based abstracting services, have led to what many regard as the regrettable demise of a number of the previously invaluable and highly acclaimed Specialist Periodical Reports of the Royal Society of Chemistry. For the past seventeen years the chapter on gas-phase photoprocesses in the SPR entitled 'Photochemistry' served as a comprehensive annual reference source for research workers interested in gas-phase spectroscopy, photochemistry, and photochemical kinetics. Alas, the explosion in the literature, brought about largely as a result of the introduction of laser light sources, meant that the contributors to recent volumes have struggled to cope with an ever-expanding subject area at a time when increased production costs were dictating a contraction in coverage. In recognition of these problems the Editors (both previous contributors to 'Photochemistry') suggested that a new informal series of review volumes be produced, designed to meet the needs of the scientific community in the fields of gas kinetics, photochemistry, gas-phase spectroscopy, and molecular dynamics, to expand upon the limited coverage offered by the single chapter and to replace, in part, the sadly missed SPR entitled 'Gas Kinetics and Energy Transfer'. This book is the first volume of our informal series, and plans for a second are in hand. The Editors welcome suggestions for future titles.

Reading M. N. R. ASHFOLD
January 1987 J. E. BAGGOTT

Contributors

P. Andresen, *Max-Planck-Institut für Strömungsforschung, Göttingen, West Germany*

R. Bersohn, *Columbia University, New York, U.S.A.*

F. F. Crim, *University of Wisconsin, Madison, Wisconsin, U.S.A.*

M. P. Docker, *University of Nottingham*

A. Hodgson, *Donnan Laboratory, University of Liverpool*

Y. T. Lee, *Lawrence Berkeley Laboratory, University of California, Berkeley, California, U.S.A.*

M. Noble, *University of Southern California, Los Angeles, California, U.S.A.*

I. Powis, *University of Nottingham*

H. Reisler, *University of Southern California, Los Angeles, California, U.S.A.*

R. Schinke, *Max-Planck-Institut für Strömungsforschung, Göttingen, West Germany*

J. P. Simons, *University of Nottingham*

C. Wittig, *University of Southern California, Los Angeles, California, U. S. A.*

A. M. Wodtke, *Lawrence Berkeley Laboratory, University of California, Berkeley, California, U.S.A.*

Contents

**Chapter 3 Dissociation of Water in the First Absorption Band:
A Model System for Direct Photodissociation**
P. ANDRESEN and R. SCHINKE

Chapter 4   **High-Resolution Photochemistry: Quantum-State Selection and Vector Correlations in Molecular Photodissociation**
M. P. DOCKER, A. HODGSON, and J. P. SIMONS

Chapter 5   **Photodissociation Processes in NO-Containing Molecules**
H. REISLER, M. NOBLE, and C. WITTIG

Chapter 6   The Dissociation Dynamics of Highly Vibrationally Excited Molecules
F. F. CRIM

Chapter 7   Characterization and Uses of Ions Generated by Multi-Photon Ionization
I. POWIS

CHAPTER 1

Photodissociation Dynamics

R. BERSOHN

This chapter aims to describe some of the major accomplishments of photodissociation dynamicists. Some important topics such as infrared multiple-photon dissociation (IRMPD), dissociation of van der Waals complexes, and fluorescence from dissociating states have been omitted, and subjects well covered in other chapters of this book are also slighted. The sections following the introduction are ordered according to the complexity of the molecule being dissociated, *i.e.* diatomics, triatomics, tetra-atomics, penta-atomics, and molecules consisting of six or more atoms.

1 Introduction

First, let us put photodissociation in its place. The photochemist is interested both in bimolecular processes involving electronically excited molecules and in unimolecular isomerizations. In his world, photodissociation plays the useful but humble role of a generator of radicals: CH_3 radicals from $Hg(CH_3)_2$, OH radicals from H_2O_2, and so forth. Inasmuch as his usual milieu is the condensed phase, even the final-state distribution of the radicals is of little interest to him. Moreover, there are no known biological processes involving photodissociation. Given this background, why is there such great interest on the part of physical chemists and molecular physicists in the photodissociation of molecules? It is true that photodissociation processes are important in the upper atmospheres of planets and the cooler regions of the stars. However, we believe that the real reasons are the appealing theoretical simplicity of the process and the availability of lasers to carry out conceptually simple experiments.

A collision between two spherically symmetric centres (for example a pair of atoms) is commonly described in terms of a scattering wavefunction that is an infinite series in the possible values of the orbital angular momentum of the colliding pair. We have no control over this angular momentum, but, except for the very long-range Coulomb potential, there is no worry over the convergence of the series. When the orbital angular momentum becomes

1

large, the particles do not approach each other closely and hence continue to move in straight lines. Stated mathematically, the corresponding phase shift is zero. As soon as there are even three atoms, as in a collision of an atom with a diatomic molecule, the picture is much more complicated.[1] The angular momentum conservation law:

$$L + j = L' + j' \tag{1}$$

for the sum of the orbital angular momentum and the rotational angular momentum of the diatomic gives a hint of the problem. As a practical matter we seldom know the initial j, but, even if a specific j and m_j have been selected, the scattering wavefunction will be an infinite sum over L, L', j', and $m_{j'}$. Although a few such calculations have been performed, they are, in general, formidable. The use of classical trajectories is a partial solution, but experiments are often carried out on specific rotational states for which detailed quantum theories are necessary.

The root of the difficulty is that we do not know the initial orbital angular momentum. Thus our ability to relate macroscopic observations on reactive scattering to microscopic events on a potential surface is limited to a considerable extent. The 'half collision' of photodissociation is an attractive alternative. The parent molecule can be cooled by jet expansion so that its rotational angular momentum is equal to or close to zero. Thus, if a molecule AB where A is an atom and B is a molecular fragment is dissociated:

$$AB + h\nu \rightarrow A + B(j) \tag{2}$$

then the angular momentum conservation law can be written:

$$J_{AB} + j_{h\nu} = j_A + L' + j \tag{3}$$

The photon adds or subtracts at most one quantum of angular momentum so that the left-hand side of the equation can often be neglected; the electronic angular momentum of the atom A is also neglected. This means that we do not have to sum separately over the quantum numbers L' and j because they must be equal to each other. The wavefunction is a sum over two quantum numbers rather than four as in the ordinary full collision.

The most attractive aspect of the half collision is that the distribution of the molecular fragment B over rotational states is also a distribution over L', which in turn is equivalent to a distribution over the exit channel impact parameter, b:

$$j \cong L' = \mu \nu b \tag{4}$$

where μ is the reduced mass of the fragments and ν is their relative speed. The spectroscopy of the molecular fragment provides an immediate physical picture of the collision dynamics. For example, when the linear molecule CS_2 is photodissociated at 193 nm, CS fragments are generated with large

[1] A. M. Arthurs and A. Dalgarno, *Proc. R. Soc. London, Ser. A*, 1960, **256**, 540.

amounts of rotational energy.[2,3] The molecule must therefore be bent in the upper state from which it dissociates. An opposite example is that of H_2S, which when cooled by supersonic jet expansion and photodissociated at 193 nm yields SH radicals that are rotationally very cold.[4] It is clear that little torque is exerted on the SH radical as it separates from the H atom.

The term 'photodissociation dynamics' means the mechanics of photodissociation, *i.e.* the forces acting during dissociation. The forces may be derived from a potential function; indeed, the ultimate aim of a study in photodissociation dynamics is the extraction of a potential surface from the experimental data. Moreover, there are very often several potential surfaces involved in the process, and one wishes to know the efficiency of surface crossing and the region in configuration space where this occurs. Lest these goals appear to be too remote or pretentious, let us consider a specific example.[5] Methylamine strongly absorbs certain lines of the CO_2 laser. The two dissociation processes requiring the least energy are a molecular elimination:

$$CH_3NH_2 \rightarrow H_2 + CH_2{=}NH \qquad (\Delta H^\circ = 32\,\text{kcal mol}^{-1})* \qquad (5)$$

and formation of a radical pair by C—N cleavage:

$$CH_3NH_2 \rightarrow CH_3 + NH_2 \qquad (\Delta H^\circ = 85\,\text{kcal mol}^{-1}) \qquad (6)$$

Xiang and Guillory have found that the second process occurs. As each i.r. quantum is equivalent in energy to $3\,\text{kcal mol}^{-1}$ and typically one quantum is absorbed per picosecond, the molecule vibrates with ever increasing amplitude for about 15—20 ps without ever finding the 'easiest' exit channel. This true dynamic constraint is easily pictured. A very high potential barrier prevents H atoms from opposite ends of the molecule from approaching each other to within a bonding distance.

While there are numerous experimental approaches to their measurement, the desired experimental quantities in photodissociation dynamics are always:

(a) the identity of the fragments,
(b) their distribution over electronic, vibrational, rotational, and translational states,
(c) the correlation, if any, of the velocities and angular momenta of the fragments with the ε vector of the polarized light that generated them, and
(d) the lifetime of the excited state.

The identity of the products is determined by mass spectrometry and spectroscopy. At sufficiently high photon energies there will be two or more sets of photofragments (open exit channels); it is often difficult to determine

---

* $\text{kcal} = 4.184\,\text{kJ}$

[2] S. C. Yang, A. Freedman, M. Kawasaki, and R. Bersohn, *J. Chem. Phys.*, 1980, **72**, 4058.
[3] J. E. Butler, W. S. Drozdoski, and J. R. McDonald, *Chem. Phys.*, 1980, **50**, 413.
[4] W. G. Hawkins and P. L. Houston, *J. Chem. Phys.*, 1982, **76**, 729.
[5] T. Xiang and W. A. Guillory, *J. Chem. Phys.*, 1986, **85**, 2019.

accurately their relative yields. State distributions were probed as long ago as the 1920s by resolving the fluorescence from excited fragments.[6] The grandchild of this technique is the detection of resolved i.r. fluorescence from vibrationally excited fragments.[7-10] Absorption has been used to probe state distributions via electronic transitions (dye lasers) and also vibrational transitions (diode lasers).[11,12] Absorption is often measured indirectly, *i.e.* by measuring intensities of the fluorescence excited by absorption or of ionization current caused by further subsequent absorption.

Laser technology now permits probing at most wavelengths. The dye lasers that emit in the region of 320—900 nm are now so powerful that their light can be efficiently frequency-doubled in crystals and acceptable intensities of vacuum ultraviolet (v.u.v.) light can be generated by four-wave mixing.[13] What does one do with molecules that do not fluoresce and have inconvenient absorptions? Coherent anti-Stokes Raman scattering (CARS), a form of resonant scattering, can be used in such cases. For example, when O_3 was dissociated by visible light into $O(^3P)$ and $O_2(X^3\Sigma_g^-)$, the O_2 was probed by the CARS technique and was shown to be strongly rotationally excited.[14] Inasmuch as the departing O atom is not the central one, the O_2 is necessarily rotated during the dissociation.

Direct and Indirect Dissociations

Photodissociations are of two types, direct and indirect. The distinction is fundamental and far-reaching. The direct dissociation takes place on the potential-energy surface of the initially excited state. Dissociation is completed in less time than a rotation period, *i.e.* about 10^{-13} s, and a large fraction of the energy release is as translational energy. The indirect dissociation takes place only after the molecule passes from the initially excited state to a lower excited state (electronic predissociation) or to the ground state (vibrational predissociation). Electronic predissociation is usually much faster than vibrational predissociation and releases much more kinetic energy. When a hot ground state dissociates, usually most of the energy release is in the form of vibrations.

[6] G. C. Neuimin in 'Elementary Photoprocesses in Molecules', ed. B. S. Neporent, Consultants Bureau, New York, 1968, p. 3.
[7] S. L. Baughcum and S. R. Leone, *J. Chem. Phys.*, 1980, **72**, 6531.
[8] S. L. Baughcum and S. R. Leone, *Chem. Phys. Lett.*, 1982, **89**, 183.
[9] J. O. Chu, G. W. Flynn, C. J. Chen, and R. M. Osgood, jun., *Chem. Phys. Lett.*, 1985, **119**, 206.
[10] S. R. Goates, J. O. Chu, and G. W. Flynn, *J. Chem. Phys.*, 1974, **81**, 4521.
[11] C. F. Wood, J. A. O'Neill, and G. W. Flynn, *Chem. Phys. Lett.*, 1984, **109**, 317.
[12] H. Kanamori, J. E. Butler, K. Kawaguchi, C. Yamada, and E. Hirota, *J. Chem. Phys.*, 1985, **83**, 611.
[13] J. W. Hepburn, *Isr. J. Chem.*, 1984, **24**, 273.
[14] D. S. Moore, D. S. Bomse, and J. J. Valentini, *J. Chem. Phys.*, 1983, **79**, 1745.

Mistakes

Experiments in photodissociation dynamics are strikingly simple in concept but in practice are sometimes difficult to execute. Even when the experiments are carried out well, mistakes in interpretation have been made. For example, atomic fragments have been found to be produced in both the ground and a low-lying metastable state. In older work the fraction of the latter was often underestimated because the quenching effect of the metastable atoms by the molecules in the system was underestimated. In several cases, techniques for measuring rotational-state distributions were used that later proved to treat different rotational states in a most unequal manner and thereby gave a false distribution. State distributions determined by laser-induced fluorescence (LIF) have been extrapolated or interpolated into regions of v and J which could not be probed by LIF. It is dangerous to do this, for some distributions are really 'surprising'. For example, the vibrational distribution of the SO product of the dissociation of SO_2 at 193 nm has been recently determined to be $N_1 = 0.2$, $N_2 = 0.7$, $N_3 = 0.03$, $N_4 = 0$, $N_5 = 0.07$ by diode laser absorption.[12] (N_0 could not be determined by this technique.) If all five populations had not been determined, the strange bimodality would have been missed.

2 Diatomic Molecules

Diatomic molecular spectroscopy reveals three types of absorption from a bound ground state: those between pairs of discrete vibration–rotation states, those in which the upper state is broadened somewhat by predissociation, and those involving a bound-to-continuum transition. In the past, little information was available on the last. Now, however, by observing the angular distribution of the atoms generated by photolysis using polarized light, the symmetry of the upper state can be determined. The distribution of the atomic fragments over the electronic states can be measured by atomic spectroscopy or by measurement of their translational-energy distribution.

The symmetry of the upper state is determined in the following way. The angular distribution of the velocities of fragments dissociated from a molecule by polarized light is given by:

$$f(\theta_{v,\varepsilon}) = [1/(4\pi)] [1 + \beta P_2(\cos\theta_{v,\varepsilon})] \qquad (7)$$

where $\theta_{v,\varepsilon}$ is the angle between the atomic velocity, v, and the ε vector of the dissociating light. For a diatomic molecule with axial recoil (dissociation fast compared to rotation) β is usually close to 2 (parallel transitions) or -1 (perpendicular transitions). Figures 1 and 2 show the classic example of the determination of the symmetry of the upper state.[15] When I_2 absorbs 465 nm light it can reach either the $B^3\Pi_0$ state or the $^1\Pi_1$ state via transitions that have parallel and perpendicular symmetry, respectively. The variation of the intensities of the two peaks as a function of angle identifies

[15] R. J. Oldman, R. K. Sander, and K. R. Wilson, *J. Chem. Phys.*, 1971, **54**, 4127.

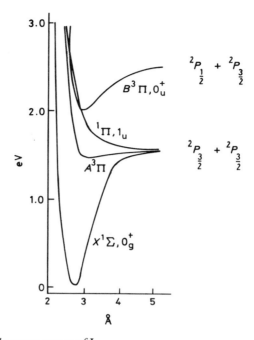

Figure 1 *Potential energy curves of* I_2
(Reproduced with permission from 'Photochemistry of Small Molecules', John Wiley, New York, 1978)

the faster and slower peaks as being due to a perpendicular ($^1\Pi_1 \leftarrow X$) and a parallel ($^3\Pi_0 \leftarrow X$) transition, respectively. The kinetic energy of the perpendicular peak corresponds to two $^2P_{\frac{3}{2}}$ I atom products whereas the kinetic energy of the parallel peak corresponds to the formation of one $^2P_{\frac{3}{2}}$ and one $^2P_{\frac{1}{2}}$ atom.

Figures 3 and 4 give an example of an accurate resolution of two overlapping continua.[16] At 305 nm KI is dissociated only into $K(^2S)$ and $I(^2P_{\frac{3}{2}})$, and at 295 nm only $K(^2S)$ and $I(^2P_{\frac{1}{2}})$ are formed. At 300 nm, which corresponds to a trough in the KI absorption, an equal mixture of the two I atom states is generated. What has been measured for I_2 and KI can be measured in principle for any diatomic. Their continuous absorptions are no longer unquantifiable.

While the translational-energy distribution of fragments is usually determined from a distribution of times of flight, Doppler spectroscopy is sometimes a convenient alternative. A fragment moving with velocity $v_z = v\cos\theta_{z,v}$ towards the observer absorbs light at a frequency $\nu = \nu_0 (1 + v\cos\theta_{z,v}/c)$, where $\theta_{z,v}$ is the angle between the velocity of the atom and the direction of propagation of the probing light. The distribution

[16] N. J. A. van Veen, M. S. de Vries, and A. E. de Vries, *Chem. Phys. Lett.*, 1979, **60**, 184.

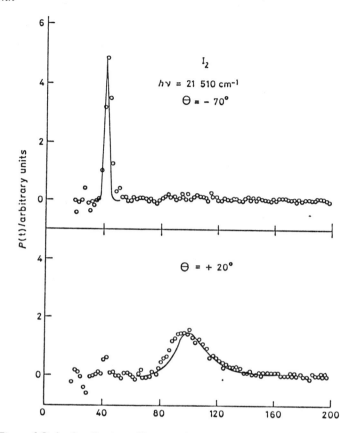

Figure 2 *Time-of-flight distribution of* I *atoms from* I_2 *photodissociation at* 465 nm. *Abscissa units are micro-seconds after the laser flash*
(Reproduced with permission from *J. Chem. Phys.*, 1971, **54**, 4127)

of velocities in the direction of the probing light is obtained from equation (7) by using the Legendre polynomial addition theorem:

$$F(\theta_{z,v}) = [1/(4\pi)]\,[1 + \beta P_2(\cos\theta_{z,\varepsilon})P_2(\cos\theta_{z,v})] \qquad (8)$$

and the lineshape has the form:

$$F(v) = [1/(4\pi)]\,[1 + \beta P_2(\cos\theta_{z,\varepsilon})\{\tfrac{3}{2}[(v - v_0)/v_0]^2(c/v)^2 - \tfrac{1}{2}\}] \qquad (9)$$

if $|v - v_0| \leqslant v_0(v/c)$. Otherwise $F(v) = 0$.

Welge and co-workers dissociated HI with polarized 266 nm light and measured the excitation spectrum of the $2p \rightarrow 1s$ fluorescence of the hydrogen atoms.[17] The spectrum was the resultant of two curves of the form of equation (9), one with $\beta = 2$, the other with $\beta = -1$; two transitions of parallel and perpendicular symmetry are implied, producing $I(^2P_{\frac{1}{2}})$ and $I(^2P_{\frac{3}{2}})$ atoms, respectively.

[17] R. Schmiedl, H. Dugan, W. Meier, and K. H. Welge, *Z. Phys. A*, 1982, **304**, 137.

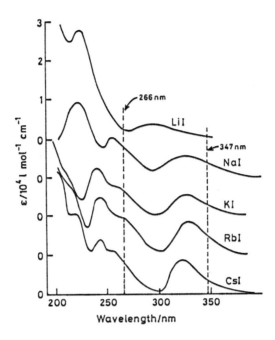

Figure 3 *Absorption spectra of the alkali iodide molecules. The spectra of the different salts are displaced from each other for convenient viewing*

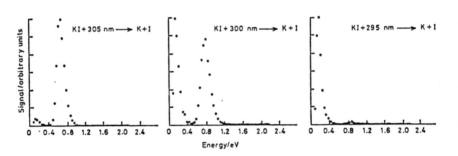

Figure 4 *Translational-energy distributions of fragments from* KI *photodissociation at* 305, 300, *and* 295 nm.
(Reproduced with permission from *Chem. Phys. Lett.*, 1979, **60**, 184)

Gerber and Moller performed an experiment, unique in many ways, on Na_2, which was excited from the $X^1\Sigma_g^+ v'' = 8$, $J'' = 42$ state to the $B^1\Pi_u$ $v' = 31$, $J' = 42$ state, from which with a lifetime of 1—2 ns it predissociated into a $3^2P_{\frac{3}{2}}$ Na atom and a $3^2S_{\frac{1}{2}}$ Na atom.[18] The former was probed by further excitation as shown in Figures 5 and 6. The Doppler-broadened

[18] G. Gerber and R. Moller, *Phys. Rev. Lett.*, 1985, **55**, 814.

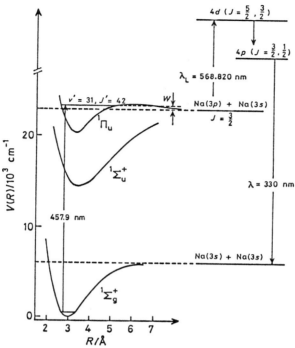

Figure 5 *Schematic representation of the excitation process and the dissociation of a quasi-bound level of the* Na_2 $^1\Pi_u$ *state. The scheme used for the detection of the atomic fragments (including observation of the* 330 nm *line) is shown on the right-hand side*

(Reproduced with permission from *Phys. Rev. Lett.*, 1985, **55**, 814)

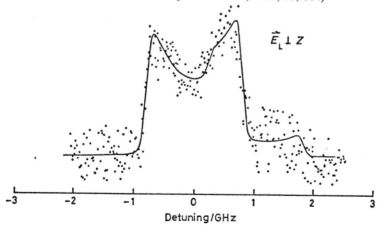

Figure 6 *Doppler spectrum of* Na^* $3^2P_{\frac{3}{2}}$, *fragments resulting from dissociation of the quasi-bound level* $v' = 31$, $J' = 42$ *of the* $B^1\Pi_u$ *state of* Na_2. *The solid curve represents a fragment distribution with* 400 cm^{-1} *kinetic energy and* $\beta = -1$ *anisotropy parameter. Two extra bumps on the right are due to the overlapping of the* $^2D_{\frac{3}{2},\frac{5}{2}} \leftarrow {}^2P_{\frac{3}{2}}$ *transitions*

(Reproduced with permission from *Phys. Rev. Lett.*, 1985, **55**, 814)

fluorescence spectrum clearly shows that the electronic transition is of the perpendicular type. At first it seems puzzling that so much anisotropy of the Na atom velocity remained after so many rotation periods of the parent Na_2. The solution to the puzzle is that for a Q-branch transition the transition dipole moment is parallel to the angular momentum of the molecule. When the excited molecule finally dissociates, the atoms will fly apart in the plane perpendicular to the transition dipole moment. This experiment is particularly important because it illustrates a method for state selection of a molecule prior to dissociation. Another remarkable method is given in Chapter 3.

3 Triatomic Molecules

The dissociation of a triatomic molecule is the theoretical heart of photodissociation dynamics. The simplest possible dissociation is the collinear dissociation of a triatomic ABC:

$$ABC + h\nu \rightarrow A + BC(\nu) \tag{10}$$

One starts in principle with ν_1 and ν_3 quanta in the stretching modes ν_1 and ν_3 of the triatomic. (The degenerate bending vibration is ignored in this collinear world.) A quantum of light is absorbed and there is now a discrete set of final states labelled by the quantum number ν of the diatomic fragment BC. In other words the asymptotic wavefunction is:

$$\Psi(R,r) = \sum_{\nu=0}^{\nu_{max}} c_\nu \exp(ik_\nu R)\psi_\nu(r) \tag{11}$$

where k_ν is the relative momentum of the system divided by $\hbar$, R is the distance between the centres of mass of the atom and the diatomic, and r is the vibrational co-ordinate of the diatomic BC. ν_{max} is the largest vibrational quantum number allowed by the conservation of energy. An experiment furnishes the populations $P_\nu = |c_\nu|^2$.

The calculation of the coefficients c_ν has been carried out exactly by Shapiro and others for such model molecules as ICN, HCN, CO_2, and N_2O.[19] Band and Freed have developed extensively a Franck–Condon model for this photodissociation.[20] The coefficients c_ν are approximated by the projection of the initial wavefunction (a product of two harmonic oscillator wavefunctions) onto the final wavefunction (a product of the harmonic oscillator wavefunction, ψ_ν, and an Airy function). The latter is the solution of the Schrödinger equation for a linear repulsive potential. The rate of transition from an initial ground state with normal co-ordinates Q_1 and Q_2 to the final state with co-ordinates R and r is proportional to:

$$\Gamma_{f\nu \leftarrow i\nu_1\nu_2} = X^2|\int S_E(R)\psi_\nu(r)\psi_{\nu'1}(Q_1)\psi_{\nu'2}(Q_2)dRdr|^2 \tag{12}$$

[19] G. G. Balint-Kurti and M. Shapiro in 'Photodissociation and Photoionization', ed. K. P. Lawley, John Wiley, 1985, p. 403.
[20] K. F. Freed and Y. B. Band in 'Excited States', ed. E. Lim, Academic Press, 1978, Vol. 3, p. 109.

where S_E is the wavefunction for translational motion at energy E and X is the transition dipole moment evaluated at a centroid of the distribution of initial co-ordinates. The desired probability is just:

$$P_v = \Gamma_{fv \leftarrow iv_1 v_2} / \sum_{v=0}^{v_{max}} \Gamma_{fv \leftarrow iv_1 v_2} \qquad (13)$$

Equation (12) is admirably simple in structure but is really valid only when the upper state, vertically above the ground state, can be thought of as a loosely coupled pair of fragments. If, instead, after the excitation the system moves (the wave packet propagates) on the upper surface, the molecule will acquire a different structure and a different set of vibrational frequencies at the transition state just before dissociation.[21-23] In this case the proper projection would be of the wavefunction at the transition state onto the fragments' wavefunctions.

In the preceding collinear model, vibrations were considered but rotations were ignored. But, according to Walsh's rules, the lower excited states of a 16-valence-electron triatomic, including the popular molecules ICN, OCS, and CS_2, and the tetra-atomic HCNO are bent; the upper-state potential will inevitably produce rotation of the diatomic fragment. Even if this were not true, the bending co-ordinates disappear and are replaced by rotational co-ordinates; the zero-point energy of the bend becomes converted into rotational energy.

The opposite physical situation in which vibrations can be ignored and the excess energy channelled into rotation and translation is often encountered. Beswick and Gelbart devised another type of Franck–Condon theory to describe this type of system.[24] They began with a Hamiltonian in which the distance between centres of mass R and the internuclear distance r were frozen at values R_0 and r_0, respectively. Their Hamiltonian function was:

$$H = \left(\frac{\hbar^2}{2\mu_{A,BC}R_0{}^2}\right)l^2 + \left(\frac{\hbar^2}{2\mu_{B,C}r_0{}^2}\right)j^2 + V(\theta) \qquad (14)$$

where l and j are the orbital angular momentum and the rotational angular momentum of the diatomic, respectively. The bond angle θ could have any equilibrium value between $0°$ and $180°$. If a total angular momentum $J = l + j$ is defined, a body-fixed representation can be chosen in which four commuting angular momentum operators are j^2, J^2, $J\cdot Z$, and $J\cdot R$, where Z is a unit vector along a space-fixed axis. The simultaneous eigenfunctions of these operators are:

$$\psi_{jJM\lambda} = [(2J+1)/4\pi]^{\frac{1}{2}} D_{M\lambda}^{J}(\varphi_R,\theta_R,0) Y_j(\theta,\varphi) \qquad (15)$$

θ_R and φ_R define the direction of R in the space-fixed system. θ and φ define the direction of the axis of the BC molecule with respect to the body-fixed system, whose z-axis is R. The bound wavefunction describing the torsional

[21] K. C. Kulander and E. J. Heller, *J. Chem. Phys.*, 1978, **69**, 2439.
[22] J. P. Simons and P. W. Tasker, *Mol. Phys.*, 1973, **26**, 1267; 1974, **27**, 1691.
[23] R. T. Pack, *J. Chem. Phys.*, 1976, **65**, 4765.
[24] J. A. Beswick and W. M. Gelbart, *J. Phys. Chem.*, 1980, **84**, 3148.

motion with quantum number n and the overall rotation with quantum numbers J and M can always be expanded in the complete set of functions $\Psi_{jJM\lambda}$:

$$|\chi_{nJM}^{bound}> = \sum_{j,\lambda} c_{j\lambda}^{nJM} \Psi_{jJM\lambda} \tag{16}$$

The sudden approximation ascribes physical reality to this expansion. The probability of producing a diatomic rotor in the state j when the parent has quantum numbers JMn is:

$$P_j^{JMn} = \sum_{\lambda=-\min.(j,J)}^{+\min.(j,J)} |c_{j\lambda}^{nJM}|^2 \tag{17}$$

When $J = 0$, this sum collapses to a single term; the simplification in the theory of photodissociation is evident.

Morse and Freed have combined the above two sudden-approximation models into a single model.[25] The essence of the model is the infinite-order sudden approximation. The dissociation is assumed to take place so quickly that the wavefunction remains invariant; the coefficients of the expansion of the initial wavefunction in terms of states of free fragments are taken to be probability amplitudes. The details of the upper-state potential are unimportant as long as it is steeply repulsive. The sudden approximation for rotation is attractive because it is so simple; if it is correct, the rotational distribution will be independent of the photon energy. The only test of this prediction known to us is for the ICN molecule. Photodissociation of ICN at increasing photon energy results in increasing average energy released in rotation.[26] Even this comparison may be inappropriate because of extensive surface crossing occurring among the excited states.

The result of a complete study of the photodissociation dynamics of a triatomic is a distribution over v and J and sometimes over other quantum numbers as well. It is only from the specific form of the final-state distribution that one can extract detailed information about the upper potential surface. Nevertheless one can average over all the possible final events and obtain average values $<f_T>$, $<f_V>$, and $<f_R>$ for the fractions of available energy released as translation, vibration, and rotation. Inasmuch as we define electronic-excitation energy to be part of the energy of dissociation, $<f_T + f_V + f_R> = 1$ and a triangular plot is useful to depict the three types of energy release in an even-handed way.[27] The data for a number of such dissociations are plotted in Figure 7. One sees at once that except for two isolated cases, H_2O at 130 nm and ClCN at 193 nm, the fraction of energy released as rotation is usually less than 0.2.[28] The molecules are divided into two classes, those for which the majority of the available energy is released as translational energy and those whose major form of release is in vibration. The first category comprises molecules of the

[25] M. D. Morse and K. F. Freed, *J. Chem. Phys.*, 1983 **78**, 6045.
[26] E. M. Goldfield, P. L. Houston, and G. S. Ezra, *J. Chem. Phys.*, 1986, **84**, 3120.
[27] R. Bersohn, *J. Phys. Chem.*, 1986, **88**, 5145.
[28] B. A. Walte and B. J. Dunlap, *J. Chem. Phys.*, 1986, **84**, 1391.

type A—B—C or A—B≡C, where the single bond is broken, and the second category those of the type A=B=C, where a double bond is broken. In the first case the electronic excitation is somewhat localized and dissociation results from a simple repulsion between A and B; in the second case the electronic excitation is delocalized, both bonds are extended before dissociation, and the BC molecule is vibrationally excited.

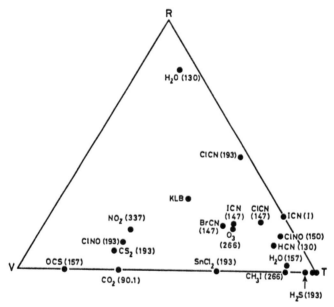

Figure 7 *A triangular plot of averaged modes of energy release of triatomic molecules dissociated at particular wavelengths. KLB is a hypothetical statistical triatomic*

4 Tetra-atomic Molecules

The tetra-atomic molecules whose photodissociation dynamics have been studied so far are divisible into two topological classes. There are molecules such as HOOH, NCCN, and HONO that have a chain-like form ABCD. There are also molecules like CH_2O and NH_3 in which one atom is bound to three others; these AX_2Y and AX_3 molecules we will call 'stars'. When a chain-like molecule ABCD with three chemical bonds dissociates, the simplest routes of dissociation would involve the breaking of one of these three bonds. In most cases it appears that, except for acetylenes, whose BC bond is a triple bond, the breaking of the BC bond is a lower-energy process than the breaking of either the AB or the CD bond (see Table 1). It is interesting that even at very high energies, when other channels such as the breaking of the AB or the CD bond are open, the dissociating molecule greatly prefers to break the weaker BC bond. There are implications here for

Table 1 *Photodissociation energies of some tetra-atomic molecules**

Chains ABCD	AB + CD	$\Delta H°$/kcal mol^{-1}	A + BCD	$\Delta H°$/kcal mol^{-1}
HOOH	2OH	50	H + O$_2$H	89
HNCO	NH($a^1\Delta$) + CO	115 ± 5	H + NCO	113 ± 6
NCCN	2CN	134	N + CCN	?
HNNN	NH($a^1\Delta$) + N$_2$	47	H + NNN	80
HONO	HO + NO	48	H + NO$_2$	78
NCNO	NO + CN	49	O + NCN	172
HCCH	2CH(X$^2\Pi$)	228	H + CCH	124

Stars AX$_2$Y and AX$_3$	AY + X$_2$	$\Delta H°$/kcal mol^{-1}	AY + 2X	$\Delta H°$/kcal mol^{-1}	AXY + X	$\Delta H°$/kcal mol^{-1}
H$_2$CO	CO + H$_2$	-0.4	CO + 2H	103	HCO + H	75
Cl$_2$CS	CS + Cl$_2$	57	CS + 2Cl	114	CSCl + Cl	64
Cl$_2$CO	CO + Cl$_2$	25	CO + 2Cl	82	COCl + Cl	77
Cl$_2$SO	SO + Cl$_2$	52	SO + 2Cl	108	SOCl + Cl	57
NH$_3$	NH($a^1\Delta$) + H$_2$	128	NH($a^1\Delta$) + 2H	195	NH$_2$ + H	102
PH$_3$	PH($a^1\Delta$) + H$_2$	71	PH($a^1\Delta$) + 2H	153	PH$_2$ + H	85

* Dissociation products formed in ground electronic states unless otherwise stated

the shape of the excited-state potential surfaces. The star-like molecules AX_2Y have two obvious pathways for dissociation, the breaking of the AX or AY bond. In reality, as seen in Table 1, the AY bond is a strong double bond. Instead of breaking just one bond, two bonds may be broken and a new one made, *i.e.* AY and X_2 molecules may be formed. This molecular elimination process (see Table 1) usually requires less energy than the radical pair formation. Both the molecular and radical pathways have been observed to occur in H_2CO, $CSCl_2$, and $SOCl_2$ at relatively low energies. At still higher energies three-body dissociation is seen in $SOCl_2$.

Table 2 *Energy partitioning in photodissociation of tetra-atomic molecules*

Chains ABCD	$<f_T>$	$<f_V>$	$<f_R>$	λ/nm	*Ref.*
HOOH	0.92	0	0.08	193	29
	0.96	0	0.04	248	
NCNO	0.31	NO: 0.09	0.29	514.6	30
		CN: 0.06	0.25		
HONO	0.77	OH: 0	0.03	369	31
		NO:			
HNNN	<0.85	NH: 0	0.15	266	32
		N_2:			
HNCO	<0.94	NH: 0	0.06	193	33
		CO:			
NCCN	0.75	0.06	0.19	193	34, 35
HCCH	0.4	0.6	0	193	36, 37
Stars AX_2Y *and* AX_3					
H_2CO	0.65	CO: 0.01	0.13	339	38
		H_2: 0.16	0.05		
$CSCl_2$	0.29			248	39
$SOCl_2$	0.61			248	40
	0.50			193	
NH_3	Small	Large		193	37
PH_3	Small	Large		193	37
AsI_3	0.26			300	41
	0.29			280	

[29] G. Ondrey, N. van Veen, and R. Bersohn, *J. Chem. Phys.*, 1983, **78**, 3732.
[30] C. Wittig, I. Nadler, H. Reisler, M. Noble, J. Catanzarite, and D. Radhakrishnan, *J. Chem. Phys.*, 1986, **85**, 1710; 1985, **83**, 5581.
[31] R. Vasudev, R. N. Zare, and R. N. Dixon, *J. Chem. Phys.*, 1984, **80**, 4863.
[32] A. P. Baronavski, R. G. Miller, and J. R. McDonald, *Chem. Phys.*, 1978, **30**, 119.
[33] W. S. Drozdoski, A. P. Baronavski, and J. R. McDonald, *Chem. Phys. Lett.*, 1979, **64**, 421.
[34] D. Eres, M. Gurnick, and J. D. McDonald, *J. Chem. Phys.*, 1984, **81**, 5552.
[35] J. B. Halpern and W. M. Jackson, *J. Phys. Chem.*, 1982, **86**, 973.
[36] A. M Wodtke and Y. T. Lee, *J. Phys. Chem.*, 1985, **89**, 4744.
[37] Z. Xu, B. Koplitz, S. Buelow, D. Baugh, and C. Wittig, *Chem. Phys. Lett.*, 1986, **127**, 534.
[38] D. Debarre, M. Lefebvre, M. Pealat, J.-P. E. Taran, D. J. Bamford, and C. B. Moore, *J. Chem. Phys.*, 1985, **83**, 4476.
[39] G. S. Ondrey and R. Bersohn, *J. Chem. Phys.*, 1983, **79**, 175.
[40] M. Kawasaki, K. Kasatani, M. Sayo, H. Shinohara, N. Nishi, H. Ohtoshi, and I. Tanaka, *Chem. Phys.*, 1984, **91**, 285.
[41] M. Kawasaki and R. Bersohn, *J. Chem. Phys.*, 1978, **68**, 2105.

The partitioning of the excess energy above the threshold energy required for dissociation has been well studied for only a few tetra-atomics at only a few wavelengths (see Table 2). The best studied molecules have been H_2CO, NCNO, NCCN, and HOOH. Each of them exhibits a different type of potential surface on which the dissociation takes place.

H_2CO has a non-degenerate $\tilde{X}^1 A_1$ ground state, as do its molecular dissociation products $H_2(X^1\Sigma_g^+)$ and $CO(X^1\Sigma^+)$, which are formed in their lowest states. As a consequence the fragments and their parent molecules correspond to different minima on the same potential surface. The process is slightly exoergic even in the absence of light; the absorption of a photon provides to the parent enough energy so that when it returns to the ground state it can overcome the approximately 80 kcal mol^{-1} barrier to dissociation. This high barrier is related to the fact that the reverse process:

$$H_2 + CO \rightarrow H_2CO \tag{18}$$

has a high barrier. The large release of translational energy (Table 2) shows that the system falls down a steeply repulsive surface. This could either be an excited surface or the ground surface with a high barrier. The correlation argument just given rules out the first possibility. Schaefer and co-workers calculated the position of the barrier in configuration space, or, in other words, the shape of the molecule in the transition state.[42] The H_2 is pushed to one side of the CO axis.

Schinke has recently computed the rotational distributions of the CO and H_2 formed in the photodissociation of H_2CO.[43] A key feature of the calculation was the choice of the initial state. It was not the ground vibrational state of the molecule of C_{2v} symmetry but rather a molecule oscillating about the unsymmetrical transition state. A sudden transition between this activated molecule and the free fragments was assumed, *i.e.* a very rapid vibrational predissociation. It turned out that, because of its almost spherical charge distribution, the rotational distribution of the H_2 molecule was determined mainly by the Franck–Condon overlap. The rotational-state distribution of the CO fragment was mainly determined by the final-state interactions. Negligible correlation was found between the H_2 and CO rotational distributions. This was a direct consequence of the fact that when the potential between the fragments was expanded:

$$V(R,\gamma_1,\gamma_2) = \sum_{\lambda_1\lambda_2} V_{\lambda_1\lambda_2}(R)P_{\lambda_1}(\cos\gamma_1)P_{\lambda_2}(\cos\gamma_2) \tag{19}$$

with $\lambda_1 = 0, 1,..., 8$ and $\lambda_2 = 0, 2$, the terms for which both λ_1 and λ_2 were not zero were negligible. The infinite-order sudden approximation was used, in which the time for dissociation on the S_0 surface is short compared to a rotation but still long compared to an H_2 vibration. Thus dissociation is sudden for the rotations but not for the vibrations.

[42] J. D. Goddard, Y. Yamaguchi, and H. F. Schaefer, tert., *J. Chem. Phys.*, 1981, **75**, 3459.
[43] R. Schinke, *Chem. Phys. Lett.*, 1985, **120**, 129; *J. Chem. Phys.*, 1986, **84**, 1487.

NCNO, nitrosyl cyanide, dissociates into $CN(X^2\Sigma^+)$ and $NO(X^2\Pi_i)$, which can be found in eight nearly degenerate states. Only one of these states can correlate with the ground state and thus there are seven excited states that correlate with the ground-state radicals. Dissociation at photon energies just above the threshold produces rotationally cold radicals, which shows that the dissociation is taking place from the ground state. The potential surface for this ground state has a very different form from that for H_2CO. The reverse reaction:

$$NC + NO \rightarrow NCNO \qquad (20)$$

has no obvious barrier and therefore there is no reason to expect a large release of kinetic energy in the dissociation. Indeed the average fraction $<f_T>$ of the available energy released as translational energy is the lowest of all the chain-like ABCD molecules listed in Table 2. A more detailed discussion of this distribution is given in Chapter 5. It is remarkable that so small a molecule behaves almost as an ideal model for unimolecular decomposition with energy being randomly distributed over a set of degenerate final states subject to a dynamical constraint. The dynamical constraint that Wittig recommends is to exclude states in which the two fragments rotate in such a way as to compensate for a large orbital angular momentum. Once these unphysical (although not outlawed by physics) states are excluded, less rotational and more vibrational excitation of the fragments is achieved.

Cyanogen, NCCN, dissociates into two $CN(X^2\Sigma^+)$ radicals, which when separated are in a four-fold degenerate state. As the two radicals approach each other, a bound singlet and a repulsive triplet are formed. The spin-orbit coupling of these light atoms is so weak that, once the molecule is placed in a lower excited singlet state, internal conversion to the ground state is much faster than intersystem crossing to the triplet state. When 193 nm light is used, there is insufficient energy to produce electronically excited CN radicals and therefore internal conversion must occur before dissociation. At 193 nm 87% of the CN is found in the $v = 0$ state and 13% in the $v = 1$ state. For each v state the partitioning of the excess energy between rotation and translation is exactly that predicted by phase space theory (PST), *i.e.* the partitioning is just proportional to the density of states for any given J. In the case of NCNO there is more vibrational excitation found experimentally than is predicted by the PST and therefore Wittig concluded that some of the rotational states should be excluded from the theory on the grounds that they were not physically reasonable.[30] For NCCN the opposite is true: there is less vibrational excitation than is predicted by PST. We are again forced to conclude that there is a dynamical constraint, this time one that reduces the probability of exciting the very stiff CN stretches.[34]

Hydrogen peroxide, HOOH, dissociates into two $OH(X^2\Pi_i)$ radicals, each of which can be in either of two two-fold degenerate states. Twelve of the combined states are triplets that can be ignored. One is the bound singlet state and the other three are repulsive states. The clearest indication that dissociation takes place from an excited state is the fact that at 248 and 193 nm more than 90% of the available energy is released as translational energy. The rotational-state distributions peak at $N = 5$, although the distribution at 193 nm is much broader. Only a modest amount of energy is released as rotational energy and virtually none in vibration. The molecule behaves essentially as a diatomic; in the excited state there is a large potential energy repelling the two centres of mass and a small one exerting torques.

A classical trajectory calculation was carried out using a potential of the form:

$$V = A\exp(-\beta R_{O-O})(1 + c\cos\varphi) \tag{21}$$

where A, β, and c are constants, R_{O-O} is the O—O distance, and φ is the dihedral angle between the two HOO planes.[44] The twisting term, c, is a consequence of the fact that the torsional potential differs in the excited state and the ground state. To fit the observed results c was taken to be 0.05 and 0.10 at 248 and 193 nm, respectively. However, the major cause of the rotational excitation is that the repulsion is between the oxygen atoms and not between the centres of mass. It is interesting that the factor $\exp(-\beta R_{O-O})$ tends to generate OH radicals with angular momenta perpendicular to each other whereas the $\cos\varphi$ term tends to make the two angular momenta antiparallel to each other. The perpendicularity is an average, not strongly correlated property whereas the antiparallelism is a result of strong correlation.

If the H_2O_2 dissociation were really 'sudden', the dipole–dipole interaction would be the major contributor to the rotational excitation. The angular momentum distribution would be bimodal, each molecule generating a radical with large angular momentum and one with small angular momentum. To fit the observed unimodal distribution over N, the dipole–dipole interaction had to be turned on slowly as the OH radicals separate. This does not necessarily mean that the OH dipole moments develop only at large distances; it might mean that the potential at short distances restricts the large correlated twists brought about by the dipole–dipole potential.

The rotational excitation of the OH radicals increases with increasing photon energy. A simple explanation is that the molecule is excited to steeper potential surfaces on which the torques are stronger. On a given potential surface the rotational distribution was found by trajectory calculations to vary very little.

In general, the determination of the correlation between the internal states of the two diatomic fragments is extremely important, particularly when both fragments are polar. McDonald has done this for NCCN by measuring simultaneously the speed, vibrational state, and rotational state of a fragment, thus fixing the energy of the other fragment.[33] While it may be very difficult to determine the correlation of the directions of the two angular momenta with respect to each other, through the use of polarized dissociating light it should be possible to determine the direction of the angular

---

[44] R. Bersohn and M. Shapiro, *J. Chem. Phys.*, 1986, **85**, 1396.

momenta of the diatomics relative to the transition dipole moment direction in the parent molecule. Hall *et al.* have recently shown that the angular momentum of the CN dissociated from ICN with polarized light exciting a parallel transition is perpendicular to the ε vector of that light.[45]

Experiments on H_2O_2 of a very different type have been carried out by Crim and co-workers. By exciting overtones of the OH stretch, enough energy is added to the molecule to cause it to dissociate into OH radicals, whose rotational distribution was measured. This is discussed in Chapter 6. Here we just mention that this is conceptually equivalent to but a vast practical improvement over IRMPD in that a known, rather than an unknown, amount of energy is added to the molecule and specifically to one type of bond.

Hydrazoic acid, HNNN, and isocyanic acid, HNCO, have been dissociated into $NH(a^1\Delta)$ and N_2 and CO fragments, respectively. The NH fragments have been found to be rotationally hot and vibrationally cold.[32,33] This result can be understood if one notes that these molecules have 16 valence electrons and are therefore isoelectronic with N_2O and CO_2, respectively. The bent upper electronic state accounts for the rotation of the NH, and it is clear that the NH stretching vibration is a spectator during dissociation. The analogy with the triatomics suggests that when the N_2 and CO molecules are probed they will be found to be vibrationally and rotationally hot.

5 Penta-atomic Molecules

At the present time a molecule with five atoms is about the most complex system whose photodissociation dynamics can be studied completely. If a penta-atomic dissociates into an atom and a tetra-atomic, the energy released in vibration can appear in any one of six modes. No spectroscopic investigation has so far measured a simultaneous distribution over six different vibrational states of product. When there is no special symmetry, the outlook is poor.

If a molecule has a symmetry axis and fragments separate in such a way that this symmetry axis is preserved, then only a_1 vibrations of the fragments can be excited. (Degenerate vibrations might be excited in pairs such that the overall symmetry was a_1, but few potentials will allow this.) When methyl iodide is dissociated in the *A* band ($^1A_1 \leftarrow {}^1A_1$), the three-fold symmetry is necessarily preserved if the products are $CH_3(^2A_2'')$ and $I^* = I(^2P_{\frac{1}{2}})$. This means that only the symmetric bend (umbrella mode) and the symmetric stretch could be excited. The frequency of the latter is so high that it is unlikely that the repulsive forces acting during the dissociation could excite it. Thus of all six modes only the symmetric bend is excited. This prediction was proven by i.r. emission and translational-energy measurements on the CH_3 photofragments.

[45] G. E. Hall, N. Sivakumar, and P. L. Houston, *J. Chem. Phys.*, 1986, **84**, 2120.

This sort of special symmetry rigorously demands both that the molecule be symmetric and that the electronic transition be of an $A_1 \leftarrow A_1$ type. Possible other examples are the dissociation of a halogen atom from C_6H_5X, an axial CO from $Fe(CO)_5$, an I atom from BI_3, and the S atom from ethylene sulphide (thi-irane), C_2H_4S. In these cases C_{3v} or C_{2v} symmetry would have to be preserved. If so, the ethylene dissociated from C_2H_4S would be vibrationally excited in at most four rather than twelve modes. This same symmetry also rules out the excitation of rotation because the repulsive force must be directed along the symmetry axis that contains the centres of mass of both fragments.

Table 3 *Energy partitioning in photodissociation of penta-atomic molecules*

Molecule	Fragments	$<f_T>$	$<f_V>$	$<f_R>$	λ/nm	Ref.
CH_3I	$CH_3 + I^*$	0.87	0.13	0	266	46, 47
		0.89	0.11	0	248	48
		0.92	0.08	0	193	49
CF_3I	$CF_3 + I^*$	0.61	0.39	0	248	50
CH_3Br	$CH_3 + Br^*$	0.94	0.06	0	222	51
		0.91	0.09	0	193	51
	$CH_3 + Br$	0.91	0.09	0	222	51
		0.90	0.10	0	193	51
CH_2I_2	$CH_2I + I$	0.13			266	52, 7
CH_2CO	$CH_2(\tilde{a}^1A_1) + CO$		CO: 0.07	0.39	308	53, 54
$HONO_2$	$HO + NO_2$		OH: 0	0.03	193	55

The methyl halide A-band transition is uniquely favourable. If only one symmetric mode is to be excited in the CH_3 radical, then it can be considered for mathematical purposes as a diatomic. Rotations are ignored and only two co-ordinates remain. For CH_3I Shapiro has found a potential that fits both the A-band absorption spectrum and vibrational distributions generated at 266 and 248 nm.[56] It has the form:

$$V(R,r) = A\exp(-aR) + B\exp(-bR + cr) + \tfrac{1}{2}k(r - r_e)^2 \quad (22)$$

[46] R. K. Sparks, K. Shobatake, L. R. Carlson, and Y. T. Lee, *J. Chem. Phys.*, 1981, **75**, 3838.
[47] H. W. Hermann and S. R. Leone, *J. Chem. Phys.*, 1982, **76**, 4766.
[48] G. N. A. van Veen, T. Baller, A. E. de Vries, and N. J. A. van Veen, *Chem. Phys.*, 1984, **87**, 405.
[49] G. N. A. van Veen, T. Baller, and A. E. de Vries, *Chem. Phys.*, 1985, **97**, 179.
[50] G. N. A. van Veen, T. Baller, A. E. de Vries, and M. Shapiro, *Chem. Phys.*, 1985, **93**, 277.
[51] G. N. A. van Veen, T. Baller, and A. E. de Vries, *Chem. Phys.*, 1985, **92**, 59.
[52] P. M. Kroger, P. C. Demou, and S. J. Riley, *J. Chem. Phys.*, 1976, **65**, 1323.
[53] D. J. Nesbitt, H. Petek, M. F. Foltz, S. V. Filseth, D. J. Bamford, and C. B. Moore, *J. Chem. Phys.*, 1985, **83**, 223.
[54] C. C. Hayden, D. M. Neumark, K. Shobatake, R. K. Sparks, and Y. T. Lee, *J. Chem. Phys.*, 1982, **76**, 3607.
[55] A. Jacobs, K. Kleinermanns, H. Kuge, and J. Wolfrum, *J. Chem. Phys.*, 1983, **79**, 3182.
[56] M. Shapiro, *J. Phys. Chem.*, 1986, **90**, 3644.

where R is the distance between centres of mass of the I atom and the CH_3 radical and r is the distance between the centres of mass of the C atom and the group of three H atoms. r_e is the equilibrium value of the height of the CH_3 pyramid; it is zero for a methyl radical whose equilibrium structure is planar but would not be zero if a similar potential were used for the dissociation of CF_3I.

The comparison between the dynamics of dissociation of CH_3I and CF_3I provides a clear confrontation of experiment with several theoretical predictions. First, let us adopt the view that breaking of the C—I bond is equivalent to cutting a string quickly, that is a vertical (frozen nuclei) transition to a pair of non-interacting fragments. If this model were true we would expect that the CH_3 radical, strongly distorted from its equilibrium shape, would depart with considerable vibrational energy. The CF_3 radical, in contrast, released in a structure close to its equilibrium geometry would depart with little internal energy. Both of these predictions are false (Table 3). These results convince us that, at least in these cases, the final-state interactions (*i.e.* the repulsive interactions in the upper state) are very important. Shapiro has shown that, qualitatively, the impulsive-spectator model explains the results. In this model only the carbon atom moves initially as a result of the repulsive force. However, in CH_3I the carbon atom has four times the mass of the three hydrogen atoms to which it is bound. Hence a force directed at the carbon atom is acting almost at the centre of mass. On the other hand, in CF_3 the carbon atom has only about one fifth of the mass of the three fluorine atoms and hence a displacement of the carbon atom is mainly a change in the relative co-ordinates and is only a small change in the position of the centre of mass. Complicating the description a bit is the fact that the symmetric C—F stretch is no longer at a high frequency and may also be excited. Clary[57] and Hennig *et al.*[58] have also constructed potential functions to describe the CF_3I dissociation.

The photodissociation of CH_2I_2 fills in our picture of the dissociating event. The very small translational-energy release has the following causes: (i) the C—I repulsive force is not directed at the centre of mass of the CH_2I and therefore a large amount of rotational energy must be generated, (ii) there is no symmetry element preserved during the dissociation, and (iii) the CH_2I radical is 'soft', *i.e.* easily deformable. I.r. is emitted over a wide range of frequencies by the hot CH_2I radicals.[7]

Ketene, CH_2CO, provides an example of the other type of fragmentation into a triatomic, CH_2, and a diatomic, CO.[53,54] Considerations of molecular spectroscopy and correlation diagrams allow one to make useful predictions. With 308 nm light only the pair of fragments $CH_2(\tilde{a}^1A_1)$ and $CO(X^1\Sigma^+)$ are energetically accessible (see Figure 8). However, these two states must correlate with the $\tilde{X}^1A_1$ state of CH_2CO. Transition to the $\tilde{a}^1A''$ state is electronically forbidden for the C_{2v} molecule, but a vibration which lowers

[57] D. C. Clary, *J. Chem. Phys.*, 1986, **84**, 4388.
[58] S. Hennig, V. Engel, and R. Schinke, *J. Chem. Phys.*, 1986, **84**, 5444.

the symmetry, *i.e.* bends the molecule, facilitates the weak transition ($\varepsilon_{max} \sim 10\,M^{-1}\,cm^{-1}$). Both the CH_2 and CO fragments are predicted to be rotationally excited; this has been confirmed for the CO fragment.

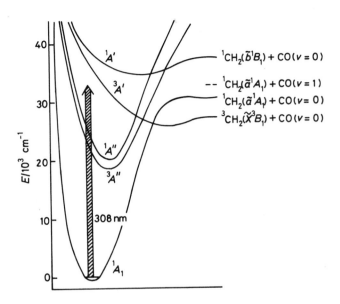

Figure 8 *Schematic potential-energy surfaces involved in near-u.v. photodissociation of ketene. The vibronically allowed excitation to the $^1A''$ state cannot directly produce $\tilde{a}\,^1A_1$ methylene; instead internal conversion to the ground state leads to dissociation without a barrier*

(Reproduced with permission from *J. Chem. Phys.*, 1985, **83**, 223)

6 Molecules with Six or More Atoms

As the number of atoms in the molecule increases, it becomes ever more difficult to probe the fragments' state distributions. Diode laser and i.r. emission spectroscopies offer great promise, however. At the present time, most of the data on photodissociation dynamics of larger molecules are on the translational energy and angular distribution of the fragments. In spite of the complexity of the molecule these two properties often provide a fairly clear picture of the photodissociation process.

Rapid Dissociations in Very Large Molecules

The most important distinction in photodissociation dynamics is not between large and small molecules but between dissociations that take place from excited states and those that result from hot ground states. To emphasize this point two examples are given here of very large molecules that dissociate on a subpicosecond time-scale.

The first example is of 1-iodonaphthalene and 2-iodonaphthalene, whose absorption spectra are both very similar to that of naphthalene and far stronger than that of CH_3I. At a wavelength of around 270 nm the π system of the aromatic rings is by far the major chromophore in the molecule. However, the π,π^* state that is excited is a delocalized state, perfectly stable as far as the C—I co-ordinate is concerned. In order for the C—I bond to break, a transition must take place to a localized state repulsive in the C—I co-ordinate. This state is probably a triplet because (i) it is repulsive and (ii) on the time-scale of a rotation (~ 1 ps) aryl iodides dissociate rapidly and aryl bromides slowly as evidenced by their velocity anisotropy and isotropy, respectively. Whatever may be its spin, the repulsive state is localized. In the same wavelength region around 270 nm, 1-bromo-, 2-bromo-, and 2-iodo-naphthalene exhibit an isotropic distribution of fragments whereas 1-iodo-naphthalene generates a strongly anisotropic distribution.[59] The bromo-substituted compounds exhibit isotropy because their electronic predissocia-tion times are longer than a rotation period. 2-Iodonaphthalene is isotropic because the direction of escape of the fragments, presumably along the C—I bond, makes a 60° angle with the transition dipole direction, the short axis of the naphthalene framework. The anisotropy parameter defined in equa-tion (7) is, if one neglects rotation before dissociation, just $2P_2(\cos\chi)$, where χ is the angle between the transition dipole moment and the dissociation direction. 60° is sufficiently close to the magic angle, 54.7°, that the distri-bution appears isotropic.

Another example is that of dirhenium (or dimanganese) decacarbonyl, $(CO)_5Re—Re(CO)_5$. This molecule falls apart into two $Re(CO)_5$ radicals when irradiated in an absorption band near 300 nm.[60] The angular distribu-tion is strongly anisotropic and the value of β shows that the transition dipole moment is along the Re—Re bond. This molecule contains 22 atoms and yet the energy is localized in one particular chemical bond and dissociation takes place on a subpicosecond time-scale. The explanation is that the electronic transition is of $\sigma^* \leftarrow \sigma$ type, in which one of the two electrons of the rhenium–rhenium bond is promoted to an antibonding state. The excited electronic state is localized and the dissociation is direct. This is not the whole story, however. The molecular-beam studies detected the dissociation:

$$(CO)_5Re—Re(CO)_5 + h\nu \rightarrow 2Re(CO)_5 \tag{23}$$

but because of a high background at mass 28 did not detect the process:

$$(CO)_5Re—Re(CO)_5 + h\nu \rightarrow (CO)_5Re—Re(CO)_4 + CO \tag{24}$$

which has been shown by other methods to be equally important.[61] In other words, internal conversion also takes place on a time-scale short compared

[59] R. Bersohn, *Isr. J. Chem.*, 1975, **14**, 111.
[60] A. Freedman and R. Bersohn, *J. Am. Chem. Soc.*, 1978, **100**, 4116.
[61] D. G. Leopold and V. Vaida, *J. Am. Chem. Soc.*, 1984, **106**, 3720.

to rotation; when it does occur, the weaker Re—CO bond is broken rather than the stronger Re—Re bond.

Three-Body Dissociations – Sequential or Simultaneous?

When a molecule dissociates into two fragments, only the momentum distribution of one fragment need be measured in order to know the total kinetic-energy distribution. The conservation of momentum supplies the necessary correlation information. When a molecule dissociates into three fragments, there is a lamentable loss of simplicity. Measurement of the momentum distribution of one fragment or even of all three does not give enough information on the correlation between the three momenta. Perhaps the most interesting aspect of three-body dissociation is the question of sequential or simultaneous bond breaking. The question is a subtle one demanding careful definition. First we consider several specific examples.

The molecule CH_3COI was photodissociated at 266 nm into CH_3, CO, and I.[62] The C—I bond is weaker than the C—C bond in the parent molecule. The I atom angular distribution was anisotropic in contrast to that of the CH_3, which was isotropic. The translational-energy release could be fitted by a model of a direct dissociation of the C—I bond with a large amount of excitation remaining in the acetyl radical, which then decomposes indirectly out of its ground state. Alas for simplicity, CF_3COI dissociation at 266 nm could not be fitted by such a model: its kinetic-energy distribution was fitted better by a simultaneous dissociation. There may be a kinematic reason for this. The centre of mass of the acetyl group of acetyl iodide lies almost on the C—I axis; the C—I repulsion will generate considerable translational energy. On the other hand, in CF_3COI the CF_3 group is so heavy relative to the carbon atom that the C—I repulsion will strongly excite the CCO bend, hastening decomposition of the CF_3CO group.

If two inequivalent bonds are broken, it is not perhaps surprising that the bond breaking is sequential with the weaker bond being broken first. What happens, however, when the two bonds are equivalent? Zewail and co-workers have recently shown that for CF_2ICF_2I direct and indirect dissociations occur in sequence.[63] When CF_2ICF_2I absorbs a 280 nm photon, it is observed that an I* atom is released in a time too short to be measured, *i.e.* less than 0.5 ps, and I atoms are released partially in an average time of 32 ± 10 ps after the initial excitation. The rapidly released I atoms probably were formerly I* atoms that changed electronic state during the rapid direct dissociation. The slowly released I atoms are the result of a unimolecular decomposition of a hot CF_2CF_2I radical. The sequence of events is exactly like that in CH_3COI with the difference that in CH_3COI the time development had to be inferred whereas in CF_2ICF_2I the time dependence is directly measured. The applications of picosecond techniques to photodissociation

[62] P. M. Kroger and S. J. Riley, *J. Chem. Phys.*, 1977, **67**, 4483; 1979, **70**, 3863.
[63] J. L. Knee, L. R. Khundkar, and A. H. Zewail, *J. Chem. Phys.*, 1985, **83**, 1896.

in the future will make a great difference to our understanding of the phenomenon.

When the quasi-linear molecule $Cd(CH_3)_2$ was dissociated with polarized u.v. light, it was found that a Cd mirror was deposited in a plane perpendicular to the ε vector of the light.[64] It follows that the momentum of the Cd atom is the sum of its original isotropic thermal momentum and a larger anisotropic momentum acquired during the dissociation. Consequently the CH_3 radicals must have unequal momenta and therefore the $Cd(CH_3)_2$ dissociation is unsymmetric. Tamir, Halavee, and Levine pointed out that, classically speaking, the truly symmetric dissociation in which the CH_3 radicals have equal and opposite momenta and the Cd atom acquires none is a unique trajectory, a set of measure zero compared to the infinite set of asymmetric trajectories.[65] Classical trajectory calculations carried out on a surface repulsive in both Cd—C co-ordinates showed that unsymmetrical dissociation is the rule; there is much more room in phase space for unsymmetrical dissociations.[65,66] The final conclusion is independent of the exact nature of the excited-state surface and details of the molecule. As long as three-body dissociation takes place on a single surface repulsive in both reaction co-ordinates, unsymmetrical but essentially simultaneous dissociations will occur.

Acetone, CH_3COCH_3, is known to dissociate at low energies within the n,π^* band into CH_3CO and CH_3 radicals; at higher energies within that same band it dissociates into two CH_3 radicals and CO. Do the two methyl radicals come off stepwise? Donaldson and Leone have excited the $^1(n,3s)$ Rydberg transition in acetone at 193 nm and found that both the non-symmetric degenerate stretching mode, v_3, of the CH_3 radical and the CO vibration are warm.[67] The CO is very strongly rotationally excited ($j_{max} \sim 32$), as is at least one of the CH_3 radicals. A possible interpretation is that the dissociation takes place by a two-step mechanism similar to the CH_3COI and CF_3COI examples discussed earlier, *i.e.* a direct dissociation from an excited surface might be followed by an indirect one on the ground-state surface. In view of the discussion on $Cd(CH_3)_2$, an alternative possibility is that the upper surface is repulsive in both Cd—C co-ordinates and dissociation is asymmetric.

The molecule *sym*-tetrazine, $C_3H_3N_3$, dissociates into three HCN molecules at 248 and 193 nm.[68] Unlike the other examples just discussed, the dissociation must take place on the ground-state surface; because HCN molecules do not easily dimerize, the dissociation is an all-or-nothing process. At 193 nm there is very little kinetic-energy release; at 248 nm there is much more. The implication is that the vibrational distribution of the HCN fragments depends on where in configuration space the system crosses

[64] J. Solomon, C. Jonah, P. Chandra, and R. Bersohn, *J. Chem. Phys.*, 1971, **55**, 1908.
[65] M. Tamir, U. Halavee, and R. D. Levine, *Chem. Phys. Lett.*,1974, **25**, 38.
[66] M. Kellman, P. Pechukas, and R. Bersohn, *Chem. Phys. Lett.*, 1981, **83**, 304.
[67] D. J. Donaldson and S. R. Leone, *J. Chem. Phys.*, 1986, **85**, 817.
[68] G. S. Ondrey and R. Bersohn, *J. Chem. Phys.*, 1984, **81**, 4517.

over to the ground state. In other words, dissociation is faster than intramolecular vibrational relaxation at 248 nm whereas at 193 nm the opposite is true.

1,2,4,5-Tetrazine, $C_2H_2N_4$, when photodissociated at 266 nm yields a fast N_2 molecule, a fast HCN molecule, and a slow HCN molecule with about 47% of the enormous available energy released as translation.[69,70] A possible mechanism is the breaking of a C—N bond followed by a ring opening and a very strong repulsion between the central HCN and its N_2 and HCN neighbours. The latter are the fast fragments and the HCN in the middle is the slow fragment.

Molecular Elimination Processes

A number of molecular elimination processes have been studied by molecular-beam methods. Examples are given in Table 4. Each of these molecular elimination processes must take place on the ground-state surface because the parent molecule and the fragments are all in non-degenerate singlet states. Basically, an activated molecule is prepared with a fixed amount of energy, E, and unimolecular decomposition takes place with a certain rate constant, $k(E)$. The angular distributions of the products of the first four reactions in Table 4 have been shown to be isotropic, but the dissociation of tetrazine is anisotropic.

Table 4   *Molecular elimination processes*

Molecule	*Fragments*	E_{AVL}/kcal mol^{-1}	$<f_T>$	λ/nm	*Ref.*
sym-Triazine	3HCN	72	0.42	248	68
		105	0.06	193	68, 10
1,2,4,5-Tetrazine	2HCN + N_2	159	0.47	266	69, 70
CH_2=CHCl	HCCH + HCl	124	0.12	193	71
cis-CHCl=CHCl	HCCCl + HCl	120	0.07	193	71
trans-CHCl=CHCl	HCCCl + HCl	120	0.10	193	71
Cyclo-octatetraene	C_6H_6 + HCCH	145	0.12	193	72
C_6H_5CH=CH_2	C_6H_6 + HCCH	109	0.12	193	72

A characteristic of these rather complex eliminations is that, in spite of the very large amount of available energy, only 10—12% is channelled into translation. (Tetrazine and *sym*-triazine at 248 nm are exceptions, perhaps because of the closed-ring structure from which their dissociations originate.) The small amount of translational energy is understood if one considers the reverse reaction. How much energy would an acetylene and a benzene molecule require in order to form styrene on collision? Obviously a

[69] D. Coulter, D. Dows, H. Reisler, and C. Wittig, *Chem. Phys.*, 1978, **83**, 1657.

[70] J. H. Glownia and S. J. Riley, *Chem. Phys. Lett.*, 1980, **71**, 429.

[71] M. Umemoto, K. Seki, M. Shinohara, U. Nagashima, N. Nishi, M. Kinoshita, and R. Shimada, *J. Chem. Phys.*, 1985, **83**, 1657.

[72] C. F. Yu, F. Youngs, R. Bersohn, and N. J. Turro, *J. Chem. Phys.*, 1985, **89**, 4409.

large amount of activation energy is required, but a collision between two fast but internally cold molecules would be non-reactive. The energy would have to be supplied primarily as energy of distortion, that is vibrational energy rather than translational energy. The consequence is that on dissociation one has relatively slow, extremely hot fragments.

Alternative Electronic Pathways to Dissociation

Hidden under a shapeless translational-energy distribution there may lie several different pathways to a given pair of dissociation products, but in general it is difficult to speculate on them. Occasionally, however, a translational-energy distribution has two peaks (*e.g.* Figures 9 and 10), and with the help of electronic spectroscopy the different indirect processes can be deduced.

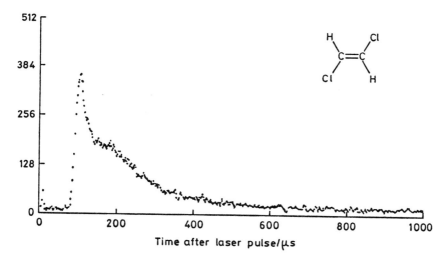

Figure 9 *Time-of-flight spectrum of* Cl *atoms dissociated from* trans-*dichloroethylene at* 193 nm
(Reproduced with permission from *J. Chem. Phys.*, 1985, **83**, 1657)

When iodobenzene was dissociated at 193 nm into phenyl radicals and iodine atoms, a fast component and a slow component were observed.[73] The fast fragments were anisotropically distributed and the slow fragments were isotropic. The interpretation is that the dissociation could take place via two different electronic states and that dissociation via the lower state took longer. For iodobenzene the specific interpretation given was that the initial state was S_3, a delocalized π,π^* state; after intersystem crossing to a repulsive triplet state localized on the C—I bond, dissociation occurred. The alternative process involved crossing from S_3 to S_1 followed by intersystem

[73] A. Freedman, S. C. Yang, M. Kawasaki, and R. Bersohn, *J. Chem. Phys.*, 1980, **72**, 1028.

crossing to the same localized repulsive triplet state. The energy difference between the S_1 and S_3 states remains in the phenyl fragment as vibrational energy. Similar observations were made on the dissociation of *trans*-dichloroethylene into ClCH=CH radicals and Cl atoms, whose translational-energy distribution very clearly reveals at least two pathways for dissociation. The specific explanation was that the molecule crosses over from its initial bound (π,π^*) state to a repulsive (π,σ^*) or (n,σ^*) state.

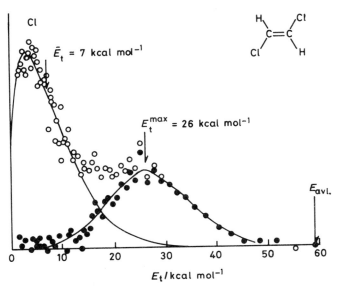

Figure 10 *Translation-energy (E_t) distribution of fragments generated by the photodissociation* ClCH=CHCl → ClCH=CH + Cl *at* 193 nm. ○: *experimental points;* ●: *high-energy component obtained by subtracting the low-energy component*
(Reproduced with permission from *J. Chem. Phys.*, 1985, **83**, 1657)

Unimolecular Dissociations

Dissociation from a hot ground electronic state is an interesting, much discussed, and still not completely understood process. The reigning theory is the Rice, Rampsberger, Kassel, and Marcus (RRKM) theory. The central assumption of the theory is that the rate of unimolecular decomposition depends only on the conserved quantities, the total energy, E, and angular momentum, J, *i.e.* there is a rate constant, $k(E,J)$, for dissociation. The molecules are assumed to be uniformly distributed over the hypersurface in phase space with energy E and angular momentum J. With pulsed laser techniques a known amount of energy can now be deposited into the ground state of a molecule either by exciting an overtone transition of a vibration or by a rapid internal conversion from an electronically excited state. Measurement of $k(E,J)$ as a function of E and J allows us to test this basic assumption of the RRKM theory and to check its more detailed predictions on the dependence of k on E and J.

Consider the dissociation of hydrogen peroxide, HOOH, following the excitation of the $\Delta v = 5$ O—H stretch overtone transition. In order for the molecule to dissociate, almost all of the energy initially localized in an O—H stretch must migrate to the O—O stretch. Scherer *et al.* have recently found that this process involves time constants of 60 and 900 ps.[74] Although the precise meaning of the two time constants has not yet been established, information on the time-scales of intramolecular energy transfer is now being acquired.

The release of a hydrogen atom from the methyl group of vibrationally excited toluene is, in principle, a similar process. In this case the photon energy (193 nm) is initially used to excite the molecule electronically. Crossing to the ground state is rapid. The hot toluene (148 kcal mol^{-1} vibrational energy) must transfer at least 88 kcal mol^{-1} to a single C—H bond in a methyl group in order for a hydrogen atom to dissociate. Dissociation is observed with a time constant of 330 ns, extracted from the rise time of laser-induced fluorescence of the hydrogen atom.[75] How fast will a similarly excited xylene (dimethyl benzene) dissociate? On the one hand, it contains two methyl groups and should therefore *a priori* dissociate twice as fast. On the other hand, a xylene has three more atoms than toluene, hence nine more vibrational modes. There are more modes over which the energy will be distributed, and it will take a longer time for sufficient energy to accumulate in one bond. In fact the hot xylenes dissociate at only one quarter of the rate of the hot toluene. The translational-energy release in toluene and the xylenes measured from the Doppler-broadened lineshape of the hydrogen atom fluorescence spectrum is only 11—13% of the available energy. Low translational energy is usually a characteristic of unimolecular dissociation from the ground state.

A caveat is necessary here. In a large molecule like toluene the return from the excited state to the ground state is finished in a time that is orders of magnitude shorter than the dissociation time. On the other hand, in a small molecule like formaldehyde the rate-determining step for dissociation may be the internal conversion. By one means or another, unimolecular decompositions can now be probed on a deeper level than ever before.

Sources. Whoever is interested in gas-phase photochemistry will find the book by H. Okabe of tremendous value.[76] Photodissociation dynamics, both the theory and the practice, have been reviewed in a number of places; progress in the field can readily be judged by the ever increasing power of the

---

[74] N. F. Scherer, F. E. Doany, A. H. Zewail, and J. W. Perry, *J. Chem. Phys.*, 1986, **84**, 1932.
[75] K. Tsukiyama and R. Bersohn, *J. Chem. Phys.*, 1987, **86**, in press.
[76] H. Okabe, 'Photochemistry of Small Molecules', John Wiley, New York, 1978.

experimental and theoretical methods reviewed there.[19,20,27,77-87] The reporter's research referred to here has been supported by the U.S. National Science Foundation and by the U.S. Department of Energy. He thanks the Chemistry Department of Tel Aviv University for hospitality while this chapter was written.

[77] R. N. Zare and D. R. Herschbach, *Proc. IEEE*, 1963, **51**, 173.
[78] R. N. Zare, *Mol. Photochem.*, 1972, **4**, 1.
[79] K. R. Wilson in 'Excited State Chemistry', ed. J. N. Pitts, Gordon and Breach, New York, 1970.
[80] R. Bersohn, *Isr. J. Chem.*, 1973, **11**, 675.
[81] J. P. Simons in 'Gas Kinetics and Energy Transfer', ed. P. G. Ashmore and R. J. Donovan (Specialist Periodical Reports), The Chemical Society, London, 1977, Vol. 2, p. 58.
[82] W. M. Gelbart, *Ann. Rev. Phys. Chem.*, 1977, **28**, 323.
[83] M. Shapiro and R. Bersohn, *Ann. Rev. Phys. Chem.*, 1982, **33**, 409.
[84] C. H. Greene and R. N. Zare, *Ann. Rev. Phys. Chem.*, 1982, **33**, 119.
[85] W. M. Jackson and H. Okabe, *Adv. Photochem.*, 1986, **13**, 1.
[86] S. R. Leone in 'Dynamics of the Excited State', ed. K. P. Lawley, *Adv. Chem. Phys.*, 1982, Vol. 50, p. 255.
[87] R. Bersohn, *IEEE J. Quant. Electr.*, 1980, **QE16**, 1208.

CHAPTER 2

High-Resolution Photofragmentation– Translational Spectroscopy

A. M. WODTKE and Y. T. LEE

1 Background Information

The advancement of lasers in combination with the molecular-beams technique has made a great impact on our understanding of primary photophysical and photochemical processes in the past twenty years. The ever increasing spectral and time resolutions, in addition to the power and range of wavelengths available, have made it possible to excite molecules selectively and with high efficiency, to study their time evolution, and to carry out state-specific detection of dissociation products. The supersonic molecular-beam source, which provides large densities of molecules with translational and rotational temperatures below a few Kelvin, has provided many new possible ways to study photochemical processes under isolated-molecule conditions.

While a number of photodissociation studies measure properties in transition between excitation and dissociation,[1] the vast majority determine so-called asymptotic properties of the dissociation process, measuring either product quantum-state distributions or velocity and angular distributions of the products. For state-specific detection of smaller fragments, especially diatomics, laser-induced fluorescence (LIF), multi-photon ionization (MPI), and coherent Raman scattering (CRS) have provided extremely detailed information on the dynamics of photodissociation.[2] Unfortunately, either

---

[1] D. Imre, J. L. Kinsey, A. Sinha, and J. Krenos, *J. Phys. Chem.*, 1984, **88**, 3956.
[2] (a) M. A. A. Clyne and I. S. McDermid, *Adv. Chem. Phys.*, 1982, **50**, 1; (b) S. R. Leone, *Adv. Chem. Phys.*, 1982, **50**, 255; (c) J. J. Valentini, 'Spectrometric Techniques', Academic Press, London, 1985, Vol. 4, p. 2; (d) D. Debarre, M. Lefebvre, M. Pealet, J.-P. E. Taran, D. J. Bamford, and C. B. Moore, *J. Chem. Phys.*, 1985, **83**, 4476; (e) C. B. Moore and J. C. Weisshaar, *Annu. Rev. Phys. Chem.*, 1983, **34**, 525; (f) D. J. Bamford, S. V. Filseth, M. F. Foltz, J. W. Hepburn, and C. B. Moore, *J. Chem. Phys.*, 1985, **82**, 3032; (g) P. Andresen, G. S. Ondrey, B. Titze, and E. W. Rothe, *J. Chem. Phys.*, 1984, **80**, 2548.

31

the vast majority of photochemically interesting product molecules cannot be detected by these methods or it is impractical to derive useful information from their spectra. The reasons for this are manifold. Consider LIF for example, where in order to determine product-state distributions quite a number of requirements must be satisfied. First, the identities of all the products of the photolysis must be well known. Second, these molecules must have optical transitions that can be efficiently probed. Third, their line strengths and transition frequencies must be well characterized. Finally, the excited state produced by the probe laser must have a fairly large quantum yield for emission of a photon as opposed to dissociation or some other dissipation process.

For most polyatomic radicals one or more of these conditions cannot be satisfied. Even when they are, because of the large excess of energy disposed into the products, a great deal more knowledge may be required than what is provided by conventional room-temperature spectroscopy. In addition, for highly internally excited molecules the inverse density of states may be much smaller than the laser bandwidth, preventing resolution and identification of the state distribution of interest.

The detection of primary dissociation products using mass spectrometers with electron impact ionization has the advantages of very high sensitivity, <1 molecule cm^{-3}, and universal detection ability. In the past the dissociation inherent in the ionization process has made it difficult to distinguish experimentally between daughter ions from electron-impact-induced dissociative ionization and photon-induced fragmentation and has limited the usefulness of mass spectrometry in the identification of primary photoproducts. An example of this can be found in the infrared multiple-photon dissociation (IRMPD) of 2-nitropropane.[3]

In this system there are two possible dissociation pathways:

$$C_3H_7NO_2 + nh\nu \rightarrow [C_3H_7NO_2]^{\ddagger} \rightarrow C_3H_7 + NO_2 \qquad (1)$$

$$C_3H_7NO_2 + nh\nu \rightarrow [C_3H_7NO_2]^{\ddagger} \rightarrow C_3H_6 + HONO \qquad (2)$$

Mass spectrometric detection of laser-dependent signals at $m/e = 46$, $NO_2{}^+$, and $m/e = 43$, $C_3H_7{}^+$, would be a good indication of the presence of channel (1); unfortunately, because of the presence of C_3H_7, which also gives $C_3H_6{}^+$ daughter ions, the detection of $m/e = 42$ would not necessarily mean that process (2) were present. Additionally, because HONO appears only as NO^+ at $m/e = 30$, a strong peak in the mass spectrum of NO_2, it would be indistinguishable by simple mass spectrometric methods.

In order to determine whether the observation of a low m/e signal is due to a neutral fragment of that mass number or merely to the ionizer-induced ionic fragment of a heavier neutral component, analytical chemists tackling the problem of complex mixture analysis often have to use a hybrid technique combining gas chromatography with mass spectrometry (GC/

³ A. M. Wodtke, E. J. Hintsa, and Y. T. Lee, *J. Phys. Chem.*, 1986, **90**, 3549.

MS), for example. It is a general feature of photodissociation that products from different decomposition pathways appear with different recoil velocities, not only governed by the interaction potential and the propensity of the system to channel energy into translation but also dependent upon the relative masses of the recoiling fragments. Therefore, if high-resolution measurement of the product velocity distribution is performed in combination with mass spectrometric detection, it is not at all essential that the photoproducts appear at their parent m/e's. This is like GC/MS on the microsecond time-scale using vacuum as the chromatography column!

In practical terms this is accomplished by producing a molecular beam in which all of the molecules of interest have approximately the same direction and velocity. By firing a pulsed laser at the beam, only dissociation products that can recoil away from the beam direction are observed in the mass spectrometric detector, which is facing the beam/laser intersection region but is situated away from the beam direction. By measuring the arrival time distribution of the neutral photoproducts over a calibrated flight length as a function of detector angle away from the beam direction, the translational-energy distribution and angular distribution of the products can be obtained.

In the IRMPD of nitropropane mentioned earlier, because of a substantial activation barrier to molecular elimination and the particular dynamics of this molecule, HONO is produced with much more translational energy than NO_2 and can be clearly resolved in the 'GC/MS', monitoring NO^+ at $m/e = 30$, shown in Figure 1d. By resolving the two components, it is possible to determine quantitatively the relative probabilities of reactions (1) and (2).

The sensitivity and resolution of the molecular-beam photofragmentation–translational–spectroscopic method, originally introduced by Wilson and co-workers,[4] have been improved immensely using second-generation molecular-beam machines in the laboratory of the Reporters over the past ten years. The determination of collision-free dissociation pathways and their relative probabilities even for quite large polyatomic molecules with complex sequential decomposition processes is a standard capability of this technique.[5] This was a critical feature of recent experiments that demonstrated bond-selective photochemistry.[6] Secondary photodissociation of primary free-radical photoproducts can also be resolved by this method and can yield interesting information on free-radical photochemistry.[7]

Because of total-energy conservation, the translational-energy distribution directly gives the product internal-energy distribution. For example,

[4] S. J. Riley and K. R. Wilson, *Faraday Discuss. Chem. Soc.*, 1972, **53**, 132.
[5] The most complex example of this is the unimolecular decomposition of $(CH_2NNO_2)_3$, known as RDX, a solid propellant and explosive. This molecule has two primary decomposition channels, each of the products of which has at least two secondary decomposition pathways. X. Zhao, E. J. Hintsa, and Y. T. Lee, to be published.
[6] L. J. Butler, E. J. Hintsa, and Y. T. Lee, *J. Chem. Phys.*, 1986, **84**, 4104.
[7] (a) A. M. Wodtke and Y. T. Lee, *J. Phys. Chem.*, 1985, **89**, 4744; (b) A. M. Wodtke, Ph.D. thesis, University of California, 1986.

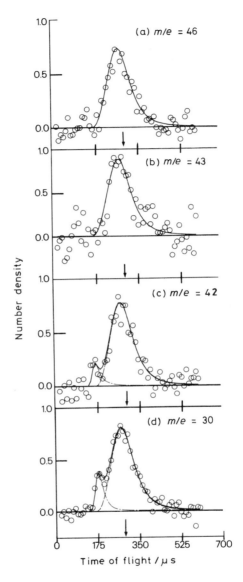

Figure 1 *TOF spectra for the IRMPD of 2-nitropropane. The laboratory angle is $10°$ from the molecular beam. The arrows indicate the beam arrival time if it were to appear at this angle. The dash-dotted line is from molecular elimination; the dashed line is from simple bond rupture. (a) $m/e = 46$, NO_2^+ from NO_2; (b) $m/e = 43$, $C_3H_7^+$ from C_3H_7; (c) $m/e = 42$, $C_3H_6^+$ from C_3H_7 and C_3H_6; (d) $m/e = 30$, NO^+ from HONO and NO_2*

(Reproduced with permission from *J. Phys. Chem.*, 1986, **90**, 3549)

in the photodissociation of O_3 [8] and CH_3I [9] the vibrational population distribution of $O_2(^1\Delta_g)$ and the CH_3 umbrella mode vibrational distribution, respectively, have been determined.

In addition, the high resolution available makes the determination of very accurate thermochemical data possible.[10] The small or non-existent absorption cross-sections at the energy threshold for dissociation of most molecules and the substantial barriers to dissociation in many molecules make it impossible to measure dissociation energies by observing the photodissociation yield as a function of wavelength. However, by photodissociating molecules well above the dissociation threshold where the absorption cross-section is substantial and by measuring the maximum release of translational energy of the products, *i.e.* the translational energy corresponding to production of ground-state products, the heats of formation of important free radicals can be obtained. Although the resolution of the translational-energy measurement is limited to ~ 1 kcal mol^{-1},* it avoids the complications of using large thermochemical cycles typical of photoionization threshold approaches that can introduce large systematic uncertainties. High-resolution time-of-flight (TOF) measurements can also yield thermochemical data on excited electronic states of free radicals that are commonly formed in photodissociation. Figure 2 shows the way in which the singlet–triplet splitting of methylene was determined from the photodissociation of ketene.[11]

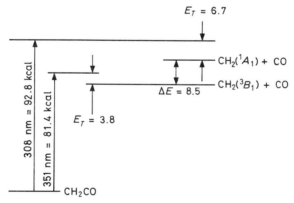

Figure 2 *The singlet–triplet splitting in methylene. Two experiments at different photon energies were performed. The two values of the maximum release of translational energy were measured, marked E_T. ΔE is the derived singlet–triplet splitting. All energies are in kcal mol^{-1}*

* 1 kcal = 4.184 kJ

[8] R. K. Sparks, L. R. Carlson, K. Shobatake, M. L. Kowalczyk, and Y. T. Lee, *J. Chem. Phys.*, 1980, **72**, 1401.

[9] R. K. Sparks, K. Shobatake, L. R. Carlson, and Y. T. Lee, *J. Chem. Phys.*, 1981, **75**, 3838.

[10] The importance of accurate thermochemical data to kinetics has been demonstrated convincingly. S. W. Benson, 'Thermochemical Kinetics', John Wiley and Sons, New York, 1976.

[11] C. C. Hayden, D. M. Neumark, K. Shobatake, R. K. Sparks, and Y. T. Lee, *J. Chem. Phys.*, 1982, **76**, 3607.

One very exciting additional fact is that the translational-energy distribution reflects the forces present *during* dissociation and as such can be used to glean very clear information on the nature of the potential-energy surface (PES). For instance, molecular elimination from formaldehyde channels a large amount of the available energy into translation whereas simple bond rupture to form HCO + H does not. This is due to a large barrier in the exit channel of the PES that efficiently channels energy into translation through a strong repulsion between the newly formed products.

The unimolecular decomposition of vibrationally excited molecules is another problem ideally suited to study by photofragmentation–translational spectroscopy. Early experiments on thermal decomposition were by necessity performed under collisional conditions since collisions provided the pumping mechanism. Today it is possible to study pyrolysis under collision-free conditions using infrared multiple photon excitation to deposit a large amount of energy into the vibrational degrees of freedom of a polyatomic molecule. Simple bond rupture reactions,[12] three- and four-centre elimination reactions,[13] as well as concerted dissociations that proceed through five- and six-membered rings have been studied systematically.

During the past two years we have constructed a new molecular-beam apparatus specifically designed and optimized for the study of photodissociation processes. This apparatus is configured with a rotating source and a fixed detector and incorporates some new ideas for background reduction and resolution enhancement. The greatly increased resolution and reduced background have enhanced our ability to study many of the questions to which we have already alluded. A few of the most recent examples are discussed in this chapter.

2 Experimental Apparatus

Figure 3 shows a detailed, scaled drawing of the instrument used in all of the experiments described below. A continuous molecular beam is produced at *1* by expanding the molecule to be studied, diluted in rare gas and typically at a pressure of 200–500 Torr* through a 125 µm nozzle. The nozzle can be heated with coaxial heating wire, shown at *2*, to increase the beam velocity and to remove clusters that can form because of the low internal temperature of the molecules in the beam. The pressure in the molecular-beam source chamber is ordinarily $\sim 10^{-4}$ Torr when the beam is running and is pumped by two 6″ diffusion pumps, one of which is shown at *10*, providing a total of $\sim 5000\,l\,s^{-1}$ pumping speed. A region of differential pumping between the source region and the main chamber helps to carry away the large gas load of the molecular-beam source so that the operating main-chamber pressure

* 1 Torr = 133.3 Pa

[12] Aa. S. Sudbø, P. A. Schultz, E. R. Grant, Y. R. Shen, and Y. T. Lee, *J. Chem. Phys.*, 1979, **70**, 912.
[13] Aa. S. Sudbø, P. A. Schultz, Y. R. Shen, and Y. T. Lee, *J. Chem. Phys.*, 1978, **69**, 2312.

is $\sim 10^{-7}$ Torr. The pressure in the main chamber is maintained by a 10″ Edwards diffstack, *9*, and by the cryopumping action of several large copper panels, *6*, cooled by liquid nitrogen. The source and differential pumping chambers are welded to a rotating vacuum seal and can themselves rotate about point *5*, where the molecular and laser beams cross. The velocity of the parent molecular beam can be measured with the retractable slotted chopping wheel shown at *8*. This entire assembly can slide downward, out of the way of the detector, without breaking vacuum when the photodissociation experiment is performed. The laser, either a Lambda-Physik EMG 103MSC excimer or a GENTEC TEA CO_2, is focused by the lens at *4* and propagates along the beam source rotation axis.

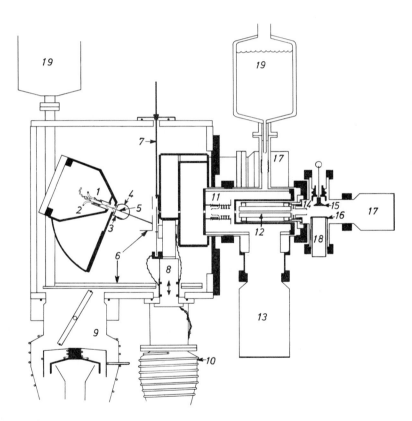

Figure 3 *The rotating-source machine. 1, molecular-beam source. 2, heating wire. 3, background 'gobbler'. 4, focusing lens for laser. 5, molecular-/laser-beam crossing region. 6, liquid-nitrogen-cooled panels. 7, gate valve assembly for detector. 8, retractable slotted chopping wheel. 9, main-chamber diffusion pump. 10, source-chamber diffusion pump. 11, Brink's-type electron impact ionizer. 12, quadrupole mass filter. 13, magnetically suspended turbomolecular pump for ionization region. 14, exit ion optics. 15, ion target. 16, scintillator. 17, grease-sealed turbomolecular pumps for differential pumping of detector. 18, photomultiplier tube. 19, liquid-nitrogen reservoirs*

The pulse of photons intersects the molecular beam at *5* and induces dissociation. A small angular fraction of the photoproducts travels through the acceptance apertures of the detector, traversing an average of 36.75 cm through two ultra-high vacuum differential pumping chambers before arriving at the ionizer. The ionization chamber is equipped with an electron impact Brink's ionizer, *11*, with an ionization length of 5 mm.[14] This gives a velocity resolution of 1.4% ($\Delta L/L = 0.5/36.75 = 0.014$). The walls of the ionization chamber are cooled with liquid nitrogen to reduce background. The ions formed at *11* are mass-analysed by an electric-quadrupole mass spectrometer at *12* and counted by a Daly-type ion counter, *15*, *16*, and *18*. The differential pumping regions of the detector and the ionization chamber are pumped by $360 \, l \, s^{-1}$ grease-sealed turbomolecular pumps, *17*, and a magnetically suspended $500 \, l \, s^{-1}$ turbomolecular pump, *13*, respectively. The absence of any lubricating fluid in the magnetically suspended pump provides a very clean vacuum.

A collimation slit at *3* cooled by liquid helium reduces background originating in the main chamber by about a factor of 10. For many experiments the limit to the background reduction possible with differential pumping is due to those molecules that pass directly through the defining slits from the main chamber to the ionizer after scattering from other background molecules or after desorbing from a surface within the viewing window of the detector. The 'background gobbler' at *3* ensures that all surfaces within the detector viewing window are at ~ 30 K. Under our experimental conditions where the main chamber is held at 10^{-7} Torr the mean free path exceeds 100 m. Almost all of the molecules that pass directly through the defining slits have to bounce off the surface in front of the detector. Cooling this surface to 30 K essentially eliminates this source of background.

The detector can be sealed off with the gate valve assembly, *7*, when the main chamber needs to be vented. The entire detector with all of the pumping equipment is mounted with ball-bearing rollers on two stainless-steel rails and can be rolled away from the main chamber under vacuum for intense baking-out. This detector design is especially effective for reducing hydrocarbon background. A partial pressure for CH_4 of 10^{-13} Torr is easily obtained in our apparatus, and with careful handling much lower partial pressures are possible.

The TOF spectrum is recorded by triggering a multi-channel scaler with the laser pulse and recording the ion counts at each m/e that shows a detectable signal as a function of arrival time. The typical time resolution of the experiment is 1—2 μs, although this can be increased to 150 ns when necessary.

---

[14] The length of the ionizer was determined by photodissociating HI at 248 nm and measuring the TOF spectrum of the H atom.

3   Laser-Selective Photochemistry in CH_2BrI^6

One of the most exciting topics in modern photochemical research is the possibility of carrying out bond-selective or mode-specific dissociation of polyatomic molecules with a laser. To succeed, two requirements must be satisfied. First, it must be possible to put the photon energy specifically into the bond that is to be broken. Second, processes that destroy the specificity of the excitation must be slow compared to the rate of dissociation. Many attempts in the past have focused on the photochemical behaviour of predissociative excited electronic states and the excitation of molecules through a specific vibrational degree of freedom by multi-photon processes. Predissociation gives rise to simple structured spectra where the first requirement above can be easily satisfied. However, it has been found essentially uniformly that energy redistribution occurs very rapidly compared to dissociation both in predissociation of excited electronic states[15] and in IRMPD.[16] This is not at all surprising since for predissociating states, by definition, energy redistribution, *e.g.* intramolecular vibrational-energy relaxation (IVR), internal conversion (IC), or intersystem crossing (ISC), must precede dissociation.

In contrast to the spectra of predissociating molecules, haloalkanes have very broad u.v. spectra because the upper electronic potential-energy surfaces are repulsive and do not support bound states. The electronic spectrum of CH_2BrI has two broad but well separated bands. One, peaking at 255 nm, involves promotion of a non-bonding electron on the I atom to an anti-bonding orbital of the C—I bond, while the other, peaking at 215 nm, is associated with excitation of a Br non-bonding electron to a C—Br anti-bonding orbital. Consequently, despite the lack of discrete structure in these spectra, specific excitation is still possible. Furthermore, since dissociation from the excited state will occur within one vibrational period, $< 10^{-13}$ s, CH_2BrI is a good candidate for bond-selective photochemistry since dissociation might be much faster than intramolecular energy transfer.

Figure 4 shows the TOF data for the photodissociation of CH_2BrI at 210 nm, exciting an electron on the C—Br bond. The solid curve that fits the

[15] (a) R. E. Smalley, *Annu. Rev. Phys. Chem.*, 1983, **34**, 129; (b) P. S. Fitch, C. A. Haynman, and D. H. Levy, *J. Chem. Phys.*, 1980, **73**, 1064; 1981, **74**, 6612; (c) C. S. Parmenter, *J. Phys. Chem.*, 1982, **86**, 1735; (d) V. E. Bondybey, *Annu. Rev. Phys. Chem.*, 1984, **35**, 591; (e) G. M. Stewart and J. D. MacDonald, *J. Chem. Phys.*, 1983, **78**, 3907; (f) R. E. Smalley, *J. Phys. Chem.*, 1982, **86**, 3504; (g) P. M. Felker and A. H. Zewail, *Chem. Phys. Lett.*, 1983, **102**, 113; (h) B. J. Van Der Meer, H. T. Jonkman, J. Kommandeur, W. L. Meerts, and W. A. Majewski, *Chem. Phys. Lett.*, 1982, **92**, 565; (i) Ph. Avouris, W. M. Gelbart, and M. A. El-Sayed, *Chem. Rev.*, 1977, **77**, 793.

[16] See for example (a) N. Bloembergen and E. Yablonovitch, *Phys. Today*, 1978, **May**, 23; (b) M. N. R. Ashfold and G. Hancock in 'Gas Kinetics and Energy Transfer', ed. P. G. Ashmore and R. J. Donovan (Specialist Periodical Reports), The Royal Society of Chemistry, London, 1981, Vol. 4, p. 73. (c) C. D. Cantrell, S. M. Freund, and J. L. Lyman in 'The Laser Handbook', ed. M. L. Stitch, North-Holland Publ. Co., 1979, p. 485; (d) P. A. Schultz, Aa. S. Sudbø, D. J. Krajnovich, H. S. Kwok, Y. R. Shen, and Y. T. Lee, *Annu. Rev. Phys. Chem.*, 1979, **30**, 379.

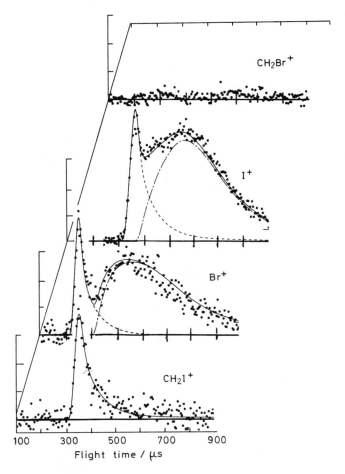

Figure 4 *TOF data from the photodissociation of* CH_2BrI. *The circles are the data and the curves are the fit to the data based on two centre-of-mass translational-energy distributions. Refer to the text for further explanation*

data in the CH_2I^+ TOF spectrum and the dashed curve that fits the fast peak in the Br^+ TOF spectrum are calculated TOF spectra based on the following reaction:

$$CH_2BrI + h\nu(210 \text{ nm}) \rightarrow CH_2I + Br \qquad (3)$$

This means that the arrival time of the CH_2I, ~ 380 µs, and the arrival time of the Br atom, ~ 280 µs, correspond to velocities in the centre-of-mass frame of reference that conserve linear momentum between the products of the above reaction. The fact that linear momentum must be balanced provides a powerful way to sort out which peaks in the spectra come from the same dissociation process, and it is much used in the analysis of TOF data. For instance, it is clear that the fast peak in the I^+ TOF spectrum is merely an electron-impact-induced fragment of the CH_2I radical since it arrives at exactly the same time as the fast peak in the CH_2I^+ TOF spectrum, ~ 380 µs. The very slow portions of the I^+ and Br^+ TOF spectra are due to three-body dissociation of CH_2BrI to $CH_2 + Br + I$, which accounts for one third of the dissociation events. This appears with much less translational-energy release since more energy is required to break two bonds. The most exciting result is that no signal appears in the CH_2Br^+ TOF spectrum, meaning that C—I bond rupture does not occur. Since the C—Br bond strength is 68 kcal mol^{-1} and the C—I bond strength is 55 kcal mol^{-1}, this experiment is the first clear example in which the laser is used to break selectively the stronger of two bonds in a molecule.

It is important to realize that the measurement of the translational energy in this experiment was critical to the successful demonstration of the presence of primary C—Br bond rupture and the absence of C—I bond rupture when CH_2BrI is excited at 210 nm. Without this knowledge there would have been no way to distinguish between C—I bond rupture, which was not occurring, and the three-body dissociation channel. Laser methods could have detected the presence of Br or I or both but they could not have made this subtle distinction.

In a similar attempt to observe bond-selective photochemistry in CF_2BrCF_2I, in which at the low temperature prevailing in the molecular beam most of the molecules should be locked into the *anti* configuration with the C—Br and C—I bonds parallel, C—Br bond excitation at 193 nm resulted in slightly more C—I than C—Br bond rupture.[17] Apparently, intramolecular electronic-energy transfer from the C—Br to the C—I bond occurs very rapidly in this molecule.

4 The Dynamics of the U.V. Photodissociation of Simple Polyatomic Molecules

By preparing molecules with a laser of narrow band width the precise starting point of a photodissociation event can be created, and by quantum-state-specific detection of the products the distribution of end-points can be determined. What happens in between is a very interesting question since photodissociation typically involves more than one PES. For small molecules the increasing accuracy of *ab initio* quantum mechanical calculations on excited-state potential-energy surfaces promises to give the most detailed information. However, for moderate- and large-sized molecules such information is not readily available.

Since the product angular and translational-energy distributions reflect the nature of the forces operating between the fragments at the critical moment of dissociation, it is possible to make inferences about the electronic

[17] D. Krajnovich, L. J. Butler, and Y. T. Lee, *J. Chem. Phys.*, 1984, **81**, 3031.

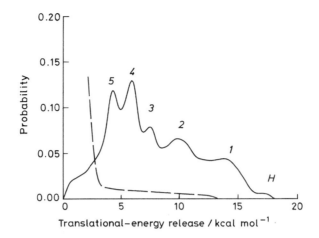

Figure 5 *Comparison of the translational-energy distributions for the photodissociation of H_2CO and C_2H_2. The solid curve is for $C_2H_2 \rightarrow C_2H + H$; the
dashed curve is for $H_2CO \rightarrow HCO + H$. The labels on the peaks of the solid
curve refer to the text*

states of the parent involved in dissociation. Figure 5 compares the translational-energy distributions of the products for photodissociation of H_2CO[18]
and C_2H_2,[7] reactions (4) and (5), both dissociating after excitation to their
lowest electronically excited singlet states:

$$H_2CO + h\nu(283.9\,\text{nm}) \rightarrow HCO + H + 14\,\text{kcal mol}^{-1} \qquad (4)$$

$$C_2H_2 + h\nu(193.3\,\text{nm}) \rightarrow HC_2 + H + 16\,\text{kcal mol}^{-1} \qquad (5)$$

The translational-energy distribution for reaction (4) is qualitatively similar
to many simple bond rupture reactions studied by IRMPD. The fact that it
peaks near zero indicates that the dissociation is from an attractive PES
without an exit barrier. It was also observed that molecular elimination
becomes less important at higher photon energy.[18] RRKM theory predicts
that reactions with small A-factors, for example molecular eliminations,
become less important at higher total energy in comparison to simple bond
rupture reactions that have large A-factors, and the experimental observation is precisely what would be expected if electronically excited formaldehyde were to undergo IC to S_0 before dissociation. This conclusion is also
consistent with a great deal of other experimental and theoretical work.[2d,2e]

In contrast, the translational-energy distribution for reaction (5) is qualitatively different. It peaks at $\sim 6\,\text{kcal mol}^{-1}$, suggesting that IC to the
ground state is unimportant and, rather, that the excited electronic state is
predissociated by either a repulsive PES or one with an exit barrier.[7] In fact,
ab initio calculations on isoelectronic HCN predict a repulsive PES that is

[18] P. Ho, D. J. Bamford, R. J. Buss, Y. T. Lee, and C. B. Moore, *J. Chem. Phys.*, 1982, **76**,
3630.

due to the interaction between the departing H atom and the forming closed-shell π orbital, all in the plane of the molecule.

In a repulsive dissociation of a rotationally cold molecule the rotational angular momentum of the products is balanced by the orbital angular momentum of the dissociation so that total angular momentum is conserved. Due to the small mass of the departing H atom and the limited available energy, only a small amount of rotational energy can be acquired by the C_2H fragment. Consequently, the vibronic structure of C_2H is partially resolved in the translational-energy distribution, and it is possible to make some very detailed conclusions about the product-energy distribution. Table 1 shows the structure of acetylene in its ground state, its electronically excited state, reached after photon absorption, and in its 'product state', $C_2H + H$, based on the most recent spectroscopic data.

Table 1 *Molecular geometry of acetylene during photodissociation*

	C—C bond length/Å	C—C—H bond angle/°	C—H bond length/Å
$C_2H_2(\tilde{X})^a$	1.2047	180	1.0585
$C_2H_2(\tilde{A})^b$	1.383	120.2	1.08
$C_2H(\tilde{X})^c$	1.2165	180	1.0464

[a]Ref. 19. [b]Ref. 20. [c]Ref. 21

While the C—C bond length and the C—C—H bending angle change substantially throughout the dissociation process, the C—H bond length changes only slightly. In fact, the changes in C—H bond length are quite a bit smaller than zero-point fluctuations. The vibrational energy is therefore expected to appear as C—C stretch and C—C—H bending vibration, and peak *1* is assigned to the vibrationless state of C_2H and peak *2* to excitation of 1 quantum of C—C stretch. The rising edge of peak *1* at 16.25 kcal mol^{-1} and that of peak *2* at 11 kcal mol^{-1} are within 100 cm^{-1} of the recently measured v_3 fundamental in C_2H.[22] Unfortunately, the bending fundamental is too low, 250 cm^{-1}, to resolve.[22] While it is clear that the C—C stretch and the C—C—H bend are highly excited in the photolysis, an unambiguous interpretation of peaks *3*, *4*, and *5* will require further theoretical and experimental effort. There appear to be two quite reasonable possibilities. C_2H has two low-lying electronic states. One possible explanation could be that both electronic states of C_2H can be formed in roughly equal amounts by the photolysis. Peaks *3*, *4*, and *5* of Figure 5 are then assigned to

[19] L. M. Sverdlov, M. A. Kovner, and E. P. Krainov, 'Vibrational Spectra of Polyatomic Molecules', John Wiley and Sons, New York, 1974, Library of Congress cat. no. QC454.S9413.

[20] C. K. Ingold and G. W. King, *J. Chem. Soc.*, 1953, 2702.

[21] Based on rotational constant for C_2H from (a) R. J. Saykally, L. Veseth, and K. M. Evenson, *J. Chem. Phys.*, 1984, **80**, 2247, and the rotational constant for C_2D from (b) M. Bogey, C. Demuynck, and J. L. Destombes, *Astron. Astrophys.*, 1985, **144**, L15.

[22] H. Kanamori, private communication.

$v = 0, 1$, and 2 of the excited-state ($^2\Pi$) bending vibration. This assignment implies that the adiabatic excitation energy is $\sim 3000 \pm 200 \, \text{cm}^{-1}$ and that the $^2\Pi$ electronic-state bending fundamental is in the neighbourhood of $550 \, \text{cm}^{-1}$.

This would appear to be the most commonplace interpretation. However, it should be stated in no uncertain terms that this is not the only possible explanation of the observed translational-energy distribution. A very interesting alternative supposes that the barrier to formation of C_2H ($^2\Pi$) in excited-state acetylene is high enough to prevent formation of the excited state. Because it is expected that, to a very good approximation, the fragmentation of excited-state acetylene occurs in a plane, dynamically it is very improbable to produce C_2H with angular momentum about the C—C internuclear axis. This is especially true for the slower part of the translational-energy distribution. Because of the cylindrical symmetry of the bending vibrational motion, only the even quanta of bending vibration can occur with vibrational angular momentum about the C—C axis equal to zero. Therefore, it is plausible that only even quanta of v_2 could be excited in the photodissociation process. Since the harmonic frequency has been found to be $250 \, \text{cm}^{-1}$ and a negative anharmonicity is expected,[23] an energy separation of $550 \, \text{cm}^{-1}$ is entirely consistent with this explanation. The disappearance of the progression at larger translational energy could be explained by the increase of product rotation at higher H atom recoil velocities or by the inherently poorer experimental resolution at larger observed laboratory velocities.

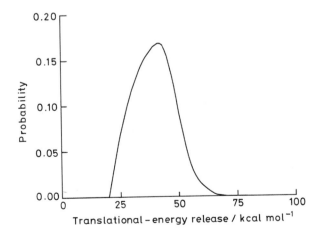

Figure 6 *The translational-energy distribution for the formation of vinyl radical from the photodissociation of vinyl bromide at 193 nm. The data from which this is derived are presented in Section 5, Figure 8*

23 W. P. Kraemer, B. O. Roos, P. R. Bunker, and P. Jensen, to be published.

In contrast to acetylene, the photodissociation of vinyl bromide, C_2H_3Br, involves substantial product rotational excitation. The translational-energy distribution for the formation of ground-state vinyl radicals at 193 nm is shown in Figure 6. Because the excited state dissociates promptly, $\sim 10^{-14}$ s, it is constrained to dissociate from geometries close to the equilibrium structure of the ground electronic state, implying a non-zero departing impact parameter. This fact, the large mass of the Br atom, the large release of translational energy, and the stiffness of the C_2H_3 radical imply the validity of a pseudo-triatomic rigid product model for a basic understanding of the dissociation dynamics. In this model energy is only allowed to appear in rotation and translation. The partitioning of energy between the two is determined by the C—C—Br angle at dissociation. The measured translational-energy distribution can be directly inverted to get the half-collision opacity function, shown in Figure 7. One can see that the opacity function is dominated by dissociation near a C—C—Br bond angle of 120°, as expected. However, because of the much smaller bending force constant of the excited PES it is possible for dissociation to occur with quite a deviation from the most favoured angle. This gives rise to the very long high-energy tail on the translational-energy distribution that gradually goes to zero as the probability for near-collinear dissociation is vanishingly small.

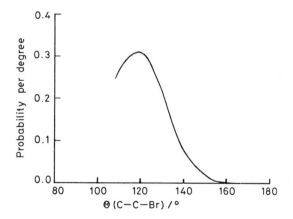

Figure 7 *The opacity function for C—Br bond rupture in the photodissociation of vinyl bromide at* 193 nm. Θ *is the C—C—Br bond angle*

5   The Determination of Free-Radical Heats of Formation

Because of their role as reaction intermediates, free radicals are one of the most interesting subjects of chemical research, and it is very important to have accurate thermochemical data about them. This is particularly true for the prediction of chemical reactivity of a radical in near-thermoneutral reactions in which a change in the heat of reaction of only a few kcal mol^{-1} can have a drastic effect on the rate. Because of the ease of detecting ions, much of present-day data is derived from threshold photoionization measurements. For example, by measuring the difference in the thresholds of reactions (6) and (7):

$$C_2HBr \xrightarrow{\text{15.56 eV}} C_2H^+ + Br + e^- \qquad (6)$$

$$C_2HBr \xrightarrow{\text{15.90 eV}} C_2H + Br^+ + e^- \qquad (7)$$

and by knowing the ionization potential (IP) for Br, 11.847 eV, the ionization potential for the C_2H radical can be obtained.[24] Combining this with the measurement of the threshold of reaction (8):[25]

$$C_2H_2 \xrightarrow{\text{17.37 eV}} C_2H^+ + H + e^- \qquad (8)$$

yields the C—H bond energy in acetylene and the heat of formation of the C_2H radical. By using high-resolution v.u.v. monochromators, a precision of one kcal mol^{-1} is obtainable.

Despite the high precision, there are a number of possible systematic difficulties with threshold measurements that give rise to large deviations between experiments, far outside the stated error bars. First of all, since it is difficult to anticipate the nature of excited states of the molecule of interest at energies as high as 15 eV, it is often an implicit assumption that ion formation is possible at the thermodynamic threshold. Problems may result from barriers to dissociation on the excited potential-energy surfaces or wavefunction symmetry restrictions that prevent the formation of ground-state products.

Secondly, ordinarily only positive ions are detected. Since most atoms and molecules have finite electron affinities, for any dissociative photoionization there will be the possibility of an ion pair production process that may form the positive ion of interest slightly below the true threshold.

Thirdly, as with any thermodynamic determination, in photoionization threshold measurements it is often necessary to rely on other thermodynamic determinations in order to build a thermochemical cycle, the missing leg of which is the process of interest. These cycles can be very complex. For example, in the above photoionization experiment the cycle begins at $C_2HBr + H$. The energy required to go to $C_2H^+ + H + Br + e^-$ is measured, process (6), and the energy released in going to $C_2H_2 + Br$ is obtained from another experiment, process (8). The well known IP of Br provides the energy required to make $C_2H_2 + Br^+ + e^-$, and the missing leg of the cycle takes us to $C_2H + H + Br^+ + e^-$. Finally, the second

[24] J. Berkowitz, 'Photoabsorption, Photoionization on Photoelectron Spectroscopy', Academic Press, New York, 1979, p. 285.
[25] V. H. Dibeler, J. A. Walker, and K. E. McCulloh, *J. Chem. Phys.*, 1973, **59**, 2264.

measured step, process (7), takes us back to the beginning. Since the energy change around the cycle is zero, the energy of the missing leg can be obtained.

While this strategy provides the experimentalist with an endless number of possible approaches to the heat of formation of a given molecule, each additional leg of a thermochemical cycle is an opportunity for the accumulation of uncertainty in the data. For the case of C_2H the problem is one of combinatorial possibilities. All of the possible combinations of the three IP determinations of C_2H, the two determinations of the threshold of process (8), and the implied acetylene C—H bond energy are tabulated in Table 2.

Table 2 *Combinatorial possibilities of the C—H bond energy in acetylene[a]*

Combination[b]	C—H bond energy/kcal mol^{-1}
DWM and WS	133 ± 1.2
DWM and MB	135 ± 1.4
DWM and OD	125 ± 1.4
ON and WS	116 ± 1.2
ON and MB	118 ± 1.8
ON and OD	107 ± 1.8

[a]This work: $D_0(C_2H—H) = 132 \pm 2$ kcal mol^{-1}. [b]DWM[25] and ON[26] are for the dissociative photoionization threshold of acetylene; WS,[27] MB,[24] and OD[28] are for the determination of the ionization potential of C_2H

In photoelectron spectroscopy this sort of difficulty has been overcome by photoionizing the parent well above threshold, where the absorption is strong, and measuring the maximum release of translational energy in the photoelectron. In this way the ionization potential can be obtained and in addition the ionization potentials for excited states of the ion can be determined in high-resolution measurements.[29] However, photoelectron spectroscopy is not as useful in the determination of dissociative ionization energies, since the translational energy of the molecular fragments is not measured.

We have been exploring the reliability of an analogous technique for the determination of bond energies. For instance, the heat of formation of C_2H can be determined by measuring the maximum release of translational energy in the photodissociation of acetylene, reaction (5). By studying the nozzle temperature dependence of the TOF spectra, the fastest C_2H, marked H in Figure 5, was determined as being due to vibrationally excited acetylene that was not fully relaxed in the expansion. Therefore the maximum release of translational energy is 16 ± 2 kcal mol^{-1}. By taking the difference

[26] Y. Ono and C. Y. Ng, *J. Chem. Phys.*, 1981, **74**, 6985.
[27] J. R. Wyatt and F. E. Stafford, *J. Phys. Chem.*, 1972, **76**, 1913.
[28] H. Okabe and V. H. Dibeler, *J. Chem. Phys.*, 1973, **59**, 2430.
[29] J. E. Pollard, D. J. Trevor, J. E. Reutt, Y. T. Lee, and D. A. Shirley, *J. Chem. Phys.*, 1984, **81**, 5302.

between this and the photon energy, 148 kcal mol^{-1}, it is easily found that the C—H bond energy in acetylene is 132 ± 2 kcal mol^{-1}.

While our result is a rigorous upper limit to the bond energy, the experiment of Abramson et al. provides a lower limit.[30] They performed anti-crossing experiments by monitoring the time-resolved fluorescence quantum beats of S_1 acetylene at a total energy of 129.5 kcal mol^{-1} and tuning triplet levels into resonance with the monitored level using a magnetic field. The large density of triplet states observed implied strong mixing of the triplet with S_0. In order for quantum beats to exist between S_1 and S_0, the bond energy must be greater than 129.5 kcal mol^{-1}.

Very recently, using a synchrotron radiation source, the threshold for proton formation from acetylene, $C_2H_2 \rightarrow C_2H + H^+$, has been measured.[31] Since the IP of H is well known, this experiment gives the most direct measurement of the C—H bond energy in acetylene by photoionization methods. The synchrotron experiments conclude that the C—H bond energy is 132 ± 2 kcal mol^{-1}, in excellent agreement with the photofragmentation work.

As has already been discussed, the assignment of the translational-energy distribution in Figure 5 is not completely unambiguous. However, it seems likely that the peaks labelled *3*, *4*, and *5* are due to the excited $^2\Pi$ electronic state of C_2H. If this is true, the data imply that the adiabatic excitation energy to the excited electronic state is 3000 cm^{-1}, the energy difference between the rising edge of peak *1* and peak *3*. This value has been particularly difficult to obtain because of extensive vibronic coupling between the two electronic states, which greatly complicates the spectrum.[32]

Another very important free radical is the vinyl radical, the heat of formation of which determines the C—H bond energy in ethylene. Over the years there have been many attempts to determine the heat of formation of the vinyl radical, the results of which vary over about 13 kcal mol^{-1}.[33] By photodissociating C_2H_3Br and measuring the maximum release of translational energy, it was hoped that a more accurate determination could be made.

The C_2H_3 radical TOF spectrum is shown in Figure 8. First of all, it is clear that there are two components. These are due to the lowest two electronic states of C_2H_3:

$$C_2H_3Br + h\nu(193.3 \text{ nm}) \rightarrow C_2H_3(\tilde{X},\tilde{A}) + Br \qquad (9)$$

[30] (a) E. H. Abramson, 'Molecular Acetylene in States of Extreme Vibrational Excitation', Ph.D. thesis, Massachusetts Institute of Technology, 1985; (b) E. H. Abramson, C. Kittrell, J. L. Kinsey, and R. W. Field, *J. Chem. Phys.*, 1982, **76**, 2293.
[31] H. Shiromaru, Y. Achiba, K. Kimura, and Y. T. Lee, *J. Phys. Chem.*, to be published.
[32] (a) P. G. Carrick, A. J. Merer, and R. F. Curl, jun., *J. Chem. Phys.*, 1983, **78**, 3652; (b) R. F. Curl, P. G. Carrick, and A. J. Merer, *J. Chem. Phys.*, 1985, **82**, 3479.
[33] (a) F. P. Lossing, *Can. J. Chem.*, 1979, **49**, 357; (b) D. J. DeFrees, R. T. McIver, and W. J. Hehre, *J. Am. Chem. Soc.*, 1980, **102**, 3334; (c) J. H. Holmes and F. P. Lossing, *Int. J. Mass Spectrom. Ion Processes*, 1984, **58**, 113; (d) A. F. Trotman-Dickenson and G. J. O. Verbeke, *J. Chem. Soc.*, 1961, 2590; (e) G. A. Chappell and H. Shaw, *J. Phys. Chem.*, 1968, **72**, 4672; (f) D. M. Golden and S. W. Benson, *Chem. Rev.*, 1969, **69**, 125; (g) G. Ayranci and M. H. Back, *Int. J. Chem. Kinet.*, 1983, **15**, 83; (h) R. B. Sharma, N. M. Semo, and W. S. Koski, *Int. J. Chem. Kinet.*, 1985, **17**, 831.

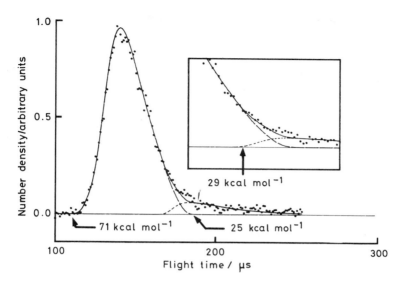

Figure 8 *The TOF distribution for the vinyl radical from the photodissociation of vinyl bromide at 193 nm. The circles are the data points. The dashed line shows the contribution from ground-electronic-state vinyl radicals (translational-energy distribution in Figure 6); the dotted line shows the contribution from the electronically excited state*

The long high-energy tail on the ground-state translational-energy distribution used to fit the fast component of the TOF spectrum is shown in Figure 6. The maximum release of translation energy, 71 kcal mol^{-1}, yields an upper limit to the heat of formation of 71 ± 3 kcal mol^{-1}.

One of the assumptions implicit in this method is that all of the available energy *can* appear as translation in some of the products. While this assumption was found empirically to be a good one in acetylene, vinyl bromide is not an ideal case and as such is a much tougher test of this assumption for several reasons. First, the excess of energy is about four times larger here than in acetylene. In addition, for a large release of translational energy, which is what we are specifically interested in, it is very improbable for an appreciable fraction of the energy not to appear as rotation. This is due to the large mass of the Br atom and to the non-zero exit impact parameter implied by the structure of the excited state. See Section 4 above.

In order to evaluate the possible error associated with the assumed formation of ground-state fragments we compared this result to a very accurate calibration experiment based on the crossed-beam reaction F + $C_2D_4 \rightarrow C_2D_3 + DF(v' = 4)$.[34] In this experiment the maximum release of the translational energy yields the heat of formation of the C_2D_3 radical,

[34] J. M. Parsons and Y. T. Lee, *J. Chem. Phys.*, 1972, **56**, 4658.

which can easily be converted to the heat of formation of C_2H_3. Figure 9 shows the angular distribution for the above reaction and the fit to the data assuming a maximum release of translational energy of 1.3 ± 0.5 kcal mol^{-1}. This particular experiment is very sensitive to the true value and yields a heat of formation for the vinyl radical of 66.7 ± 0.5 kcal mol^{-1}, slightly lower than the photodissociation experiment, as expected. The fact that it is only 4.3 kcal mol^{-1} lower than the photodissociation result gives very valuable information on the validity of the assumption that ground-state fragments are produced. As expected, it is important that the departing atom is light, that the excess of energy is not too large, and that we should be suspicious of our ability to derive the maximum translational energy when the translational-energy distribution tails away very gradually.

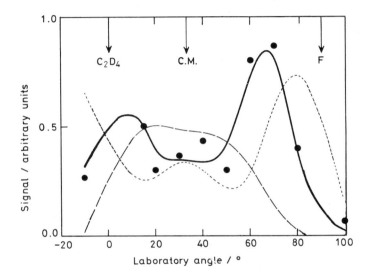

Figure 9 *Angular distribution of the* DF *product from the reaction* F + $C_2D_4 \rightarrow C_2D_3 +$ DF($v' = 4$). *The circles are the data points. The solid line shows the best fit* $P(E_T)$ *assuming* 1.3 kcal mol^{-1} *maximum translational energy, the long-dashed line assumes 0.8 kcal mol^{-1}, and the short-dashed line assumes* 1.8 kcal mol^{-1}

Turning to the translational-energy distribution for the excited-state radical, a careful inspection of the inset of Figure 8 shows a pronounced, discontinuous change in the slope of the data at a translational-energy release of 29 kcal mol^{-1}. Since there is about half as much excess of energy, even if the same proportion of rotational excitation prevents the formation of ground-state fragments, the error should be substantially smaller than the 4.3 kcal mol^{-1} found for the ground state. These data yield a heat of formation for the electronically excited vinyl radical of 113 ± 1.5 kcal mol^{-1}.

Recently, the $\tilde{A} \leftarrow \tilde{X}$ absorption spectrum was measured by Hunzicker, and a C—C stretching progression was observed.[35] However, an absolute assignment of the lines was impossible since small Franck–Condon factors prevented the observation of the electronic origin. By using the derived $T_{00}(\tilde{A}-\tilde{X})$, $46 \pm 1.5 \, \text{kcal mol}^{-1}$, we were able to make the assignment of their spectrum shown in Table 3. If this assignment were correct, the extrapolated $T_{00}(\tilde{A}-\tilde{X})$ from reference 35 would be $46.9 \, \text{kcal mol}^{-1}$.

Table 3 *Assignment of the vibronic transitions of the vinyl radical*

Observed transition[a]/cm^{-1}	Assignment[b] $m \leftarrow 0$
(16 414)	$0 \leftarrow 0$
(17 616)	$1 \leftarrow 0$
(18 818)	$2 \leftarrow 0$
20 020	$3 \leftarrow 0$
21 222	$4 \leftarrow 0$
22 427	$5 \leftarrow 0$
23 629	$6 \leftarrow 0$
24 815	$7 \leftarrow 0$
25 981	$8 \leftarrow 0$
27 137	$9 \leftarrow 0$

[a]Ref. 35; values in parentheses are extrapolated. [b]Excitation from $v = 0$ of the ground electronic state to $v = m$ of the C—C stretch in the excited electronic state

Since the photodissociation determination was fairly accurate, even in a very unfavourable case, it appears that the method of photofragment spectroscopy has a bright future with regard to the determination of thermochemical quantities. The success of this method hinges on its simplicity, since the thermochemical cycles employed in these experiments could not be smaller. Moreover, this is one of the very few ways of deriving accurate thermodynamic data for excited electronic states of free radicals.

In addition to the method of measuring the maximum release of translational energy, the C—X bond energies in quite a number of halogen-containing free radicals have been obtained by a somewhat different approach, using the photofragmentation technique. By photodissociating dihalogenated alkanes, for example CH_2ClCH_2I, at 248 nm the C—I bond is broken and vibrationally excited CH_2ClCH_2 radicals are formed with a distribution of internal energies.[36] Radicals observed at short arrival times corresponding to a large release of translational energy are the coldest internally and can be easily detected. By determining the minimum translational-energy release, *i.e.* the maximum internal energy, at which the free radical can be detected, the energy required to remove both halogens from the dihalogenated alkane is obtained. Since C—I bond energies are fairly

[35] H. E. Hunzicker, H. Kneppe, A. D. McLean, P. Siegbahn, and H. R. Wendt, *Can. J. Chem.*, 1983, **61**, 993.
[36] T. K. Minton, P. Felder, R. J. Brudzynski, and Y. T. Lee, *J. Chem. Phys.*, 1984, **81**, 1759.

similar from one molecule to another, the C—Cl bond energy in the free radical is obtained. The results for a number of such molecules are given in Table 4.[37]

Table 4   *Carbon–halogen bond energies in a number of halogen-containing free radicals*

Free radical	C—X bond energy[a]/kcal mol⁻¹
CH_2CH_2—Cl	19.5 ± 1
CF_2CH_2—Br	11.7 ± 2
CF_2CF_2—Br	21.3 ± 1
CF_2CF_2—I	8.1 ± 1.5

[a]Refer to ref. 37 for details

6   Collision-Free Pyrolysis Experiments

The discovery of collision-free IRMPD was first thought to signal the dawning of a new age of photochemistry, where chemists would be able to tailor reactions based on the ability to deposit energy specifically and directly into a chosen vibrational degree of freedom. This hope rested upon the assumption that IVR would be slower than the infrared pumping rate and the dissociation rate. It later became fully understood that this assumption was wrong by several orders of magnitude for molecules excited to near their dissociation limits. That is, under favourable circumstances infrared photons could be deposited at a rate of 10^9 s⁻¹, but for polyatomic molecules near their dissociation limit IVR proceeds on a picosecond time-scale.

Still, the death knell of mode-specific chemistry was the harbinger of a greatly increased understanding of unimolecular decomposition. The application of statistical theories of dissociation rates, especially RRKM theory, provided a quantitative connection between thermodynamics and kinetics and could also predict the product-energy distributions for many reactions. For example, by knowing the total internal energy, RRKM theory can accurately calculate the translational-energy distribution, $P(E_T)$, of simple bond rupture reactions with no exit barrier in the PES. The dynamics of concerted molecular elimination reactions cannot, however, be understood with RRKM theory. Because of the substantial exit barrier in the PES, a great deal of the available energy gets channelled into translation at the expense of vibration.

Because of picosecond IVR, the internal distribution of vibrational energy created by IRMPD is not substantially different from that produced in collisional pyrolysis experiments. So, by using a CO_2 laser in a molecular beam, it is possible to 'heat' a molecule in a collision-free environment. This allows the unambiguous determination of primary dissociation mechanisms.

[37] T. K. Minton, Ph.D. thesis, University of California, 1986.

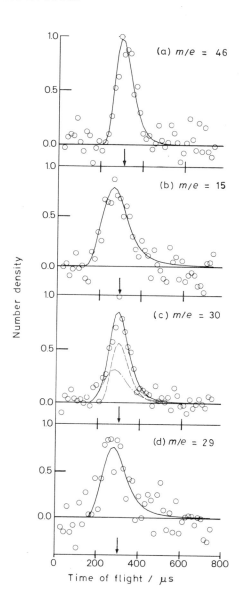

Figure 10 *TOF data from the IRMPD of* CH_3NO_2. *The laboratory angle is* $10°$ *from the molecular beam. The arrows indicate the beam arrival if it were to appear at this angle. The dash-dotted line is from isomerization; the dashed line is from simple bond rupture. The circles are the data and the curves are the fit to the data based on two* $P(E_T)s$. (a) $m/e = 46$, NO_2^+ *from* NO_2; (b) $m/e = 15$, CH_3^+ *from* CH_3; (c) $m/e = 30$, NO^+ *from* NO_2 *and* NO; (d) $m/e = 29$, HCO^+ *from* CH_3O*

(Reproduced with permission from *J. Phys. Chem.*, 1986, **90**, 3549)

For instance, reaction (10) has always been assumed to be the only primary dissociation channel in the thermal decomposition of nitromethane:

$$CH_3NO_2 \xrightarrow{\Delta} CH_3 + NO_2 \qquad (10)$$

The TOF spectra for the collision-free thermal decomposition of nitromethane, studied by molecular-beam IRMPD, is shown in Figure 10. Signals are observed at $m/e = 46, 30$, and 15, as expected from reaction (10). In addition, however, a substantial signal at $m/e = 29$ indicates the formation of the methoxy radical [reaction (11)]:

$$CH_3NO_2 + nh\nu \rightarrow [CH_3NO_2]^{\ddagger} \rightarrow [CH_3ONO]^{\ddagger} \rightarrow CH_3O + NO \quad (11)$$

Because of the lower endothermicity of reaction (11), the signal at $m/e = 29$ appears with slightly more translational energy. By careful inspection of the $m/e = 30$ TOF spectrum, it is clear that the reaction partner NO is also appearing. The branching ratio between (10) and (11) can be calculated from the ionization cross-sections, fragmentation patterns, and Jacobian transformation factors between the laboratory frame and the centre-of-mass frame. The result of such analysis shows that 38% of the dissociation events follow channel (11) under our experimental conditions. This was the first observation of the unimolecular isomerization of CH_3NO_2 into $CH_3O + NO$.[38]

These results point out how useful qualitative information about the thermal decomposition of polyatomic molecules can be obtained from experiments performed under collision-free conditions. Such information would be very difficult to obtain by other methods owing to fast secondary reactions of the primary reaction products. In addition, because of the ability to determine accurate product branching ratios and translational-energy distributions, it is possible to derive quantitative kinetic parameters. This is the subject of the next subsection.

The Derivation of Activation Energies from IRMPD Experiments

By knowing the A-factors for decomposition of the energized CH_3NO_2 molecule into both of its possible product channels, RRKM theory can be used to predict the product branching ratio as a function of the relative heights of the barrier to isomerization and the C—N bond energy at a given total excitation energy. RRKM theory can be used to calculate the $P(E_T)$ of simple bond rupture reactions without exit barriers, given the total energy. Thus one can work backwards to obtain information on the excitation-energy distribution of the dissociating ensemble of molecules by finding the excitation-energy distribution that gives a good RRKM predicted fit to the observed simple bond rupture $P(E_T)$. In our experimental arrangement, the

[38] A. M. Wodtke, E. J. Hintsa, and Y. T. Lee, *J. Chem. Phys.*, 1986, **84**, 1044.

excitation-energy distribution will be dependent upon the characteristics of the laser radiation. That is, if the laser intensity is low, the molecules will dissociate with only a small amount of excess of energy. On the other hand, if the laser intensity is high, the molecules will on average be pumped up quite high above the dissociation limit before the RRKM dissociation rates become as fast as the up-pumping rate.

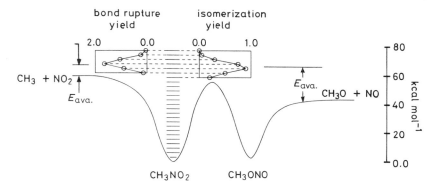

Figure 11 *Schematic representation of the barrier height determination for the isomerization of* CH_3NO_2. *The levels shown are separated by the photon energy of the laser. By adjusting the absorption cross-section of the molecule in the quasi-continuum and knowing the rate of decomposition as a function of internal excitation from RRKM theory, the bond rupture yield curve is adjusted to fit the observed translational-energy distribution and the isomerization barrier height is adjusted to fit the observed branching ratio between the two reaction channels. See ref. 3 for details*

(Reproduced with permission from *J. Phys. Chem.*, 1986, **90**, 3549)

Figure 11 shows a schematic representation of a calculation designed to determine the barrier height to isomerization in reaction (11). The dynamic analysis consists of using a master rate equation simulation program to determine the magnitude of the absorption cross-sections in the quasi-continuum that give rise to a simple bond rupture yield curve, which will reproduce the experimental translational-energy distribution for process (10). This places constraints on and gives a rough estimate of the excitation distribution in the decomposing parent molecule. Then, by varying the barrier height to isomerization, it is possible to fit the experimental branching ratio. The rate constants as a function of excitation energy are determined using RRKM theory and the theoretical A-factor of Dewar *et al.* for isomerization.[39] This branching-ratio matching method yields a nitromethane isomerization barrier height of 55.5 kcal mol^{-1}.[3] The barrier height determination can be converted to an activation energy using RRKM theory, and this is usually about a 1 kcal mol^{-1} correction.

[39] M. J. S. Dewar, J. P. Ritchie, and J. Alster, *J. Org. Chem.*, 1985, **50**, 1031.

We have also tested the branching-ratio matching method on two known decomposition systems, reactions (12) and (13):

$$C_2H_5NO_2 + nh\nu \rightarrow [C_2H_5NO_2]^{\ddagger} \rightarrow C_2H_4 + HONO \qquad (12)$$

$$CH_3CO_2C_2H_5 + nh\nu \rightarrow [CH_3CO_2C_2H_5]^{\ddagger} \rightarrow CH_3CO_2H + C_2H_4 \quad (13)$$

In both cases we have obtained excellent agreement, within the $\pm$ 1.5 kcal mol^{-1} error of our determination, with known activation energies. Some of the derived barrier heights and activation energies are listed in Table 5.

Table 5   *Activation energies and barrier heights derived from IRMPD experiments*

Reaction	Barrier height/kcal mol^{-1}	Activation energy/kcal mol^{-1}
$CH_3NO_2 \rightarrow CH_3ONO^a$	55.5	54.0
$C_2H_5NO_2 \rightarrow C_2H_4 + HONO^a$	46.0	45.0
$2\text{-}C_3H_7NO_2 \rightarrow C_3H_6 + HONO^a$	41.0	40.0
$CH_3CO_2C_2H_5 \rightarrow C_2H_4 + CH_3CO_2H^b$	50.0	49.0
$CH_3CO_2CH_3 \rightarrow CH_2CO + CH_3OH^b$	69.0	68.0

[a]Ref. 3. [b]Ref. 40

The Position of the Transition State along the Reaction Co-ordinate

As mentioned earlier, RRKM theory cannot predict the translational-energy distribution for concerted molecular elimination reactions with high exit potential barriers. In order to understand the dynamics of this kind of unimolecular decomposition it is necessary to perform accurate calculations of the multi-dimensional PES and carry out classical trajectory dynamics calculations. Alternatively, experiments are performed. There have now been quite a number of experimental measurements of translational-energy distributions for molecular elimination reactions. It is possible to compare these results and obtain a qualitative understanding of some of the aspects of these types of reactions.

In the past, several four-centre HCl elimination reactions have been studied.[13] The results of these experiments are presented in Table 6. Typically, these reactions release 15—25% of the exit barrier into translation, indicating that the products are left with much vibrational energy. This is indicative of the strained nature of the four-membered cyclic transition state. It is very interesting to ask the following question: if one studies molecular elimination reactions proceeding through five- and six-centre cyclic transition states, does the percentage of the barrier height going into translation increase, because of the smaller amount of strain present in the ring?

40 E. J. Hintsa, A. M. Wodtke, and Y. T. Lee, to be published.

Table 6 *Translational-energy release for the four-centre molecular elimination of* HCl

Reaction	Exit barrier[a] height/kcal mol^{-1}	Fraction of barrier[b] in translation
$CH_3CF_2Cl \rightarrow CH_2CF_2 + HCl$	55	0.22
$CH_3CCl_3 \rightarrow CH_2CCl_2 + HCl$	42	0.19

[a]The exit barrier is defined as the energy difference between the transition state and the products. [b]This is defined as the most probable translational-energy release divided by the exit barrier height

Table 7 shows the results of several experiments performed recently for five- and six-centre cyclic transition states. For all of these sytems at least 65% of the exit barrier appears as translation, confirming that ring size is indeed a very important factor to consider in these types of reactions.

Table 7 *Translational-energy release for five- and six-centre molecular elimination reactions*

Reaction	Exit barrier[a] height/kcal mol^{-1}	Fraction of barrier[a] in translation
$C_2H_5NO_2 \rightarrow C_2H_4 + HONO$[b,c]	28	0.71
$2\text{-}C_3H_7NO_2 \rightarrow C_3H_6 + HONO$[b,c]	23	0.65
$CH_3CO_2C_2H_5 \rightarrow CH_3CO_2H + C_2H_4$[d,e]	38	0.57
$C_2H_3OC_2H_5 \rightarrow CH_3CHO + C_2H_4$[d,f]	38	0.63

[a]Same as Table 6. [b]Five-centre elimination. [c]Ref. 3. [d]Six-centre elimination. [e]Ref. 40. [f]Ref. 41

However, in principle the dynamics should also depend on the electronic structure of the transition state. That is, if the transition state at the top of the mechanical barrier has a delocalized, hybrid electronic structure between products and reactants, more energy would be expected to go into internal degrees of freedom, since the barrier is substantially made up of potential energy due to bond extension. On the other hand, if the electronic structure of the barrier looks like the products, much energy will appear as translation, since the major contributor to the barrier is energy due to the repulsion of two closed-shell molecules being held at bonding distances from one another. This question is equivalent to asking if the barrier to molecular elimination is a late one or an early one.

Light is shed on this issue of reaction co-ordinate position of the transition state by considering the results of three four-centre elimination reactions:

$$CH_3CO_2CH_3 + nh\nu \rightarrow [CH_3CO_2CH_3]^\ddagger \rightarrow CH_2CO + CH_3OH \quad (14)$$

$$CH_3CO_2H + nh\nu \rightarrow [CH_3CO_2H]^\ddagger \rightarrow CH_2CO + H_2O \quad (15)$$

$$C_2H_5OC_2H_5 + nh\nu \rightarrow [C_2H_5OC_2H_5]^\ddagger \rightarrow C_2H_5OH + C_2H_4 \quad (16)$$

[41] F. Huisken, D. Krajnovich, Z. Zhang, Y. R. Shen, and Y. T. Lee, *J. Chem. Phys.*, 1983, **78**, 3806.

Table 8   *Translational-energy release for the four-centre molecular elimination of ROH*

Reaction	Exit barrier[a] height/kcal mol^{-1}	Fraction of barrier[b] in translation
$CH_3CO_2CH_3 \rightarrow CH_2CO + CH_3OH$[c]	34	0.62
$CH_3CO_2H \rightarrow CH_2CO + H_2O$[c]	36	0.64
$C_2H_5OC_2H_5 \rightarrow C_2H_5OH + C_2H_4$[d]	49	0.53

[a]The exit barrier is defined as the energy difference between the transition state and the products. [b]This is defined as the most probable translational energy divided by the exit barrier height. [c]Ref. 40. [d]Ref. 42

We have performed IRMPD experiments on the above four-centre elimination reactions; Table 8 shows the results. Reactions (14)—(16) release substantially more of the exit barrier energy into translation, > 50% of the barrier compared to 20% for HCl elimination. It appears that in the case of four-centre elimination of alcohols and water the transition state at the top of the barrier has an electronic structure much closer to products than does the transition state for HCl elimination. These experiments, perhaps more than any of the others described, show how the translational-energy distribution can give very detailed information on the interfragment potential-energy surface.

7   Conclusions

The purpose of this article has been to show the various *kinds* of information that can be obtained from the application of high-resolution photofragmentation–translational spectroscopy, by presenting a few examples of the most recently accomplished experiments. In summary, a novel approach to bond-selective photochemistry has been described, some aspects of u.v. photodissociation dynamics have been investigated, the derivation of accurate thermochemical data for free radicals has been presented, and the study of collision-free pyrolysis has been examined. The role of this method in the overall field of photochemistry is a unique one owing to the general nature of the detection method, which allows the study of an essentially unlimited range of polyatomic molecules. We can certainly look forward to substantial advancement of this technique in the future. In particular, there is nothing in principle that limits the improvement of the resolution of this method by up to a factor of five or more. This would allow a new generation of extremely high-resolution experiments to be performed. It is also anticipated that, with the increasing availability and reliability of short-wavelength lasers, *e.g.* the F_2 laser at 157 nm, the understanding of hydrocarbon photochemistry will be greatly improved. The usefulness of the molecular-beam technique to prepare reactive species, such as free radicals, in a collision-free environment

⁴² L. J. Butler, R. J. Buss, R. J. Brudzynski, and Y. T. Lee, *J. Phys. Chem.*, 1983, **87**, 5106.

has yet to be fully exploited, and it is expected that the study of radical photochemistry will blossom soon. In addition, the role of collision-free isomerization in u.v. photochemistry can be studied using isotopic labelling techniques. The investigation of these as well as many other topics offers a bright future for researchers using this technique.

Acknowledgement. We would like to acknowledge the efforts of Dr. Douglas J. Krajnovich and Dr. Laurie J. Butler, who both participated in the construction of the apparatus used in these experiments. In addition, the support of the Office of Naval Research and the U.S. Department of Energy over the years has been invaluable.

CHAPTER 3

Dissociation of Water in the First Absorption Band: A Model System for Direct Photodissociation

P. ANDRESEN and R. SCHINKE

1 Introduction

Photodissociation is the fragmentation of molecules due to the absorption of light. It is not only a very elementary process in the interaction of light with matter but also of considerable importance in different areas of chemistry and physics.[1] Photodissociation processes can be used to pump lasers[2] or masers[3] or even to produce chemicals.[4] It is the first step in the whole field of photochemistry, generating radicals that initiate subsequent reactions. From current experimentation it has been found that the photodissociation of H_2O in the first absorption band explains the astronomical OH masers in regions of star formation.[3] There can be no doubt that characterization of the basic principles of such processes is important.

Although it has been studied for a long time, it is only recently that photodissociation has become a very active field of research.[5] This is essentially due to the progress in laser technology that has made experiments much easier and has allowed much more sophisticated studies. The formation of the nascent products in different quantum states can be determined

[1] H. Okabe, 'Photochemistry of Small Molecules', Wiley Interscience, New York, 1978.
[2] J. V. V. Kasper and G. C. Pimental, *Appl. Phys. Lett.*, 1964, **5**, 231; J. V. V. Kasper and G. C. Pimental, *Phys. Rev. Lett.*, 1965, **14**, 352.
[3] P. Andresen, G. S. Ondrey, and B. Titze, *Phys. Rev. Lett.*, 1983, **50**, 486.
[4] J. Wolfrum, M. Kneba, P. Clough, and M. Schneider, Offenlegungsschrift DE2938353, Deutsches Patentamt, München, 2.4.1981.
[5] S. R. Leone, *Adv. Chem. Phys.*, 1982, **50**, 255; J. P. Simons, *J. Phys. Chem.*, 1984, **88**, 1287; R. Bersohn, *J. Phys. Chem.*, 1984, **88**, 5145.

by very sensitive and state-selective detection methods like laser-induced fluorescence (LIF)[6] or resonance-enhanced multi-photon ionization (REMPI).[7] Powerful lasers allow efficient excitation of parent molecules with monochromatic, and in some wavelength ranges also tunable, light. State-to-state experiments are now possible with new laser methods that allow the preparation of parent molecules in single quantum states. The potential power of these methods is demonstrated in this chapter, in which we discuss the first state-to-state experiment.[8] These techniques yield much more detailed information about the dynamics of simple photodissociation processes and lead to a much better understanding of the primary step in photochemistry.

At the same time theory has reached a state in which simple fragmentation processes can be treated without severe approximations on an *ab initio* basis, at least for a few select cases.[9] This leads to a situation in which a quantitative comparison of *ab initio* calculations with highly sophisticated experimental results becomes possible for the first time. The photodissociation of water in the first absorption band is one such case:

$$H_2O(\tilde{X}^1A_1) + h\nu \rightarrow H_2O(\tilde{A}^1B_1) \rightarrow H(^2S) + OH(^2\Pi) \qquad (1)$$

A particularly clear physical situation is found here, because only one excited-state potential surface is involved. H_2O is small enough for *ab initio* calculations to be reliable, and the corresponding potential-energy surface has been determined.[10] Due to a fortunate decoupling of rotational and vibrational degrees of freedom, it is possible to treat the dynamics in the excited state without unreasonable approximations. Many features that are quantitatively understood here are important in other processes as well, where they may also be qualitatively understood in view of these results. *This is why we consider the photodissociation of* H_2O *in the first absorption band an ideal model system for direct dissociation processes.*

Moreover, process (1) is of considerable astrophysical interest. It explains the OH maser radiation from regions of star formation, *i.e.* the pump mechanism for the *interstellar* OH masers.[11,12] Because these masers operate between the Λ-doublet states of OH, processes that yield a selective population of the upper Λ-doublet state explain the inversion between the maser levels. Indeed, the photodissociation of H_2O in the first absorption band yields OH preferentially in the upper Λ-doublet states.[3] In the course of this work it was shown that most OH maser theories were wrong, simply because they were based on an erroneous assignment of maser levels. A considerable

[6] U. Hefter and K. Bergmann in 'Atomic and Molecular Beam Methods', ed. G. Scoles, Oxford, New York, 1986.
[7] M. N. R. Ashfold, *Mol. Phys.*, 1986, **58**, 1; D. H. Parker in 'Ultrasensitive Laser Spectroscopy', ed. D. S. Kliger, Academic Press, New York, 1983, p. 233.
[8] P. Andresen, V. Beushausen, D. Häusler, and H. W. Lülf, *J. Chem. Phys.*, 1985, **83**, 1429.
[9] E. Segev and M. Shapiro, *J. Chem. Phys.*, 1982, **77**, 5604.
[10] V. Staemmler and A. Palma, *Chem. Phys.*, 1985, **93**, 63.
[11] P. Andresen, *Astron. Astrophys.*, 1986, **154**, 42.
[12] P. Andresen, *Comments At. Mol. Phys.*, 1986, **18**, 1.

part of these very detailed studies was actually motivated by these astrophysical applications: only the quantitative understanding of process (1) makes a detailed modelling of these maser sources possible.

The selective population of Λ-doublet states found in this example is not only interesting for astrophysical applications. It is found in reactive and inelastic collisions as well and usually indicates the symmetry of the transition state.[13] In the present case the origin for the Λ-doublet selectivity can be easily understood both qualitatively and quantitatively.

Although the photodissociation of H_2O has been considered in many experiments,[14] only a few studies were related to the first absorption band. The total-absorption cross-section has been measured several times,[15] the most recent study showing some structured absorption.[16] A first determination of the OH product-state distribution was reported in 1967.[17] These data indicated a 'cold' rotational distribution, in qualitative agreement with the present experiments.

In this review only some of the interesting results that were obtained in a series of recent experimental and theoretical studies will be discussed. Technical details can be found in the original publications. [3,8,13,18−25]

In the first experiment OH product-state distributions were measured following photodissociation of room-temperature H_2O at 157 nm.[18] Later on, similar experiments were performed for jet-cooled species in a nozzle beam at the same wavelength.[19] Polarization experiments were carried out subsequently for both room-temperature and jet-cooled species.[20] Finally, in the most recent experiments the photodissociation of H_2O was studied at $\lambda = 193$ nm from single rotational states in vibrationally excited H_2O.[8,21] This is the first and so far the only experiment in which a direct dissociation is studied on a state-to-state basis.

The most important experimental findings for the different degrees of freedom can be summarized as described below.

(a) *Rotation:* In all cases only a very small amount (1—2%) of the total available energy is channelled into rotation. For both the room-temperature and the jet-cooled species the rotational-state distributions can be described (with one exception) as a Boltzmann distribution. The OH rotational-energy content depends considerably upon the temperature of the parent H_2O. In contrast, the product-state distributions originating from single rotational states of H_2O show very different behaviour. They are strongly structured, are certainly non-Boltzmann, and depend sensitively on the initial rotational state.[21]

[13] P. Andresen and E. W. Rothe, *J. Chem. Phys.*, 1985, **82**, 3634.
[14] O. Dutuit, A. Tabche-Fouhaile, H. Frohlich, P. M. Guyon and I. Nenner, *J. Chem. Phys.*, 1985, **83**, 584 and references therein.
[15] K. Watanabe and M. Zelikoff, *J. Opt. Soc. Am.*, 1953, **43**, 753.
[16] H. T. Wang, W. S. Felps, and S. P. McGlynn, *J. Chem. Phys.*, 1977, **67**, 2614.
[17] K. H. Welge and F. Stuhl. *J. Chem. Phys.*, 1967, **46**, 2440.
[18] P. Andresen and E. W. Rothe, *Chem. Phys. Lett.*, 1982, **86**, 270.
[19] P. Andresen, G. S. Ondrey, B. Titze, and E. W. Rothe, *J. Chem. Phys.*, 1984, **80**, 2548.
[20] P. Andresen and E. W. Rothe, *J. Chem. Phys.*, 1983, **78**, 989.
[21] P. Andresen, D. Häusler, and R. Schinke, *J. Chem. Phys.*, to be published.

(*b*) *Vibration:* At 157 nm considerable population is found in vibrationally excited states. For all different H_2O temperatures the rotational distributions in $n = 0$ and $n = 1$ are almost identical, suggesting a decoupling of vibrational and rotational degrees of freedom. At 193 nm no vibrationally excited OH is found,[22] indicating that the vibrational distribution depends upon photolysis wavelength.

(*c*) *Spin Population:* Neither in the room-temperature nor in the jet-cooled experiments could a clear preference for one or the other multiplet state be observed. Even the state-to-state experiments indicate a statistical population of spin states. However, small non-statistical effects cannot be excluded with the present signal-to-noise ratio.

(*d*) Λ-*Doublets:* The OH product-state distributions reveal a strong propensity for the more asymmetric Λ-doublet state, especially for the jet-cooled species. Although the preference depends strongly upon the temperature of the parent molecule and the final OH rotational state, it is always for the same asymmetric Λ-doublet state. In true state-to-state experiments this is different: no consistent preference is found. Instead, either the one or the other Λ-doublet state dominates, depending both on the initial H_2O and on the final OH quantum state.[21]

The theoretical studies are based on the *ab initio* calculations of the excited-state potential surface $\tilde{A}^1B_1$.[10] Preliminary dynamical studies demonstrated that final-state interaction for the rotational degree of freedom is extremely small for this surface.[23] In particular, it turned out that the rotational and vibrational degrees of freedom decouple, *i.e.* the rotational-state distributions are independent of the vibrational motion. Because of the extremely small amount of rotational-energy transfer, the rotational degree of freedom can be approximately treated within the energy sudden limit. This in turn allows full three-dimensional calculations for the vibrational-state distributions and the total-absorption cross-section, including both dissociation channels.[24] This is the first real *ab initio* treatment for direct photodissociation of a polyatomic molecule.

In most dissociation processes the product-state distributions depend on both the excitation and the fragmentation step. In general, it is impossible to study the features of these two steps separately. Photodissociation of H_2O in the first absorption band, however, is a limiting case of a dissociation in which the rotational and electronic fine-structure distributions are determined exclusively in the first step. It offers the unique opportunity to study the characteristic features of the first step in photodissociation. The experimental and theoretical results demonstrate that these features are by no means trivial. Because this point is also important in other dissociation processes, it is stressed considerably in this chapter.

The product-state distributions for rotational and electronic fine-structure

[22] D. Häusler, thesis, Max-Planck-Institut für Strömungsforschung, Göttingen, West Germany, 1985, Bericht 25.

[23] R. Schinke, V. Engel, and V. Staemmler, *Chem. Phys. Lett.*, 1985, **116**, 165.

[24] V. Engel, R. Schinke, and V. Staemmler, *Chem. Phys. Lett.*, 1986, **130**, 413.

degrees of freedom result exclusively from the first step, because the final-state interaction is very small for these degrees of freedom. The reason for the small final-state interaction is a small angular anisotropy of the excited-state potential surface around the ground-state equilibrium geometry.[10] Therefore, the rotational degree of freedom can be accurately described in the Franck–Condon (FC) limit.[25] *The origin of the validity of the FC limit for process* (1) *results from a characteristic feature of the excited-state potential surface $\tilde{A}^1B_1$.*

Recently Balint-Kurti incorporated the electronic structure of the products OH and H into a formal theory designed especially for the description of process (1).[26] The FC limit of this theory enabled us to predict *for the first time* the population of rotational, Λ-doublet, and spin states with moderate numerical effort.[27]

These very detailed studies allow a comparison of experiment and theory on a state-to-state level. The rotational and electronic fine-structure distributions almost agree quantitatively with experiment on the basis of six (!) quantum numbers. Quantitative agreement with experiment is also obtained for the calculated total-absorption cross-section, and even finer details such as the progression of 'vibrational structures'[16] are well reproduced.[24] It will be demonstrated later on in this chapter that the photodissociation of H_2O is almost completely understood on the basis of a known potential-energy surface for the excited state and rigorous dynamical calculations including all degrees of freedom. A complete description of the process is obtained since we are now able to predict state-to-state cross-sections for *all initial H_2O and final OH quantum states* (excluding nuclear spin and magnetic quantum numbers) and for *all photolysis wavelengths*. This makes the dissociation of water in the first absorption band a unique example of direct photofragmentation.

In the next section of this chapter some general aspects of state-to-state experiments and their relation to other experiments, in which averaging over initial or final quantum states is involved in one way or the other, are examined. In Section 3 particular attention is given to the question of whether and, if so, when initial quantum states are important for direct dissociation processes. The close relation of the validity of the FC limit with the final-state interaction is considered for the dissociation of a model triatomic molecule, including only the rotational degree of freedom, and the important consequences of the FC limit are elucidated. In Section 4 the general physics of the photodissociation of H_2O in the first absorption band and the characteristic features of the *ab initio* potential-energy surface are described. In Section 5 experimental and theoretical results concerning rotational and electronic fine-structure distributions (Λ-doublet and spin) are compared with each other. The absorption cross-section and the wave-

[25] R. Schinke, V. Engel, and V. Staemmler, *J. Chem. Phys.*, 1985, **83**, 4522.
[26] G. G. Balint-Kurti, *J. Chem. Phys.*, 1986, **84**, 4443.
[27] R. Schinke, V. Engel, P. Andresen, D. Häusler, and G. G. Balint-Kurti, *Phys. Rev. Lett.*, 1985, **55**, 1180.

length-dependent vibrational distribution are presented in Section 6. The influence of averaging over an initial distribution of H_2O rotational states and of summation over OH Λ-doublet states is discussed in Section 7, and a short discussion about the astrophysical importance of the photodissociation of H_2O in the first absorption band is given in Section 8. A summary with the main conclusions of this work in Section 9 closes the chapter.

2   State-to-State Photodissociation

Both in this section and in the next some fundamental aspects of state-to-state photodissociation that are crucial for the photodissociation of water in the first absorption band are discussed. The role of the initial quantum state of the parent molecule is explained and the relationship between state-to-state cross-sections, temperature-dependent product-state distributions and the total-absorption cross-section is examined. Because these are questions of more general interest, consider the photodissociation of a model ABC molecule:

$$ABC(i) + h\nu(\lambda) \xrightarrow{\ 1\ } ABC^* \xrightarrow{\ 2\ } A + BC(f) \tag{2}$$

Here, i indicates the initial quantum state of the parent molecule ABC, $h\nu$ is the photon energy, ABC^* denotes the 'transition state' or the 'excited complex', and f represents the final quantum states in which the product BC is formed. The probability for this process, that is the probability to absorb a photon and to obtain the product BC in quantum state f if the parent was in quantum state i, is essentially the state-to-state photodissociation cross-section σ_{if}. In quantum theory this state-to-state photodissociation cross-section is formally given by the Golden Rule expression:[28]

$$\sigma_{if}(\lambda) \sim \nu \, | <\Psi^i_{gr} | \mu | \Psi^f_{ex}> |^2 \tag{3}$$

Here, Ψ^i_{gr} represents the bound-state wavefunction for the parent molecule in quantum state i and Ψ^f_{ex} is the dissociation wavefunction that asymptotically relates to the final quantum state f.

From equation (2) it is obvious that the state-to-state photodissociation cross-sections depend only on two parameters: the initial state i of the parent molecule and the energy $h\nu$ (or the wavelength λ) of the photon. All features of the products (for example the type of products, whether the products are electronically excited or in the ground state, how the products are distributed over quantum states) are determined by these two parameters alone. Whereas there is no doubt that the photon energy is very important in photodissociation experiments,[1] little is known about the role of the initial state i.

This leads to one of the major topics of this chapter: the extent to which state-to-state cross-sections depend on the initial state of the parent molecule. This is closely related to the strength of the final-state interaction in equation (2) and is discussed in the next section.

[28] M. Shapiro and R. Bersohn, *Annu. Rev. Phys. Chem.*, 1982, **33**, 409.

Averaging over State-to-State Cross-sections

State-to-state photodissociation cross-sections are measured only in very few cases. *Product-state distributions, which are usually measured, represent true state-to-state cross-sections only if photolysis occurs from a single quantum state i in the parent molecule.* In most experiments, however, photodissociation is studied in a gas at a particular temperature with the parent molecules distributed over different initial (in most cases rotational) states. The distribution of initial states at temperature T is usually given by the Boltzmann expression:

$$b_i(T) \approx (2J_i + 1)\exp(-\Delta E_i/kT) \qquad (4)$$

Under such conditions dissociation occurs from many different initial states of the parent molecule. *The product-state distribution $p_f(\lambda|T)$ now depends on temperature but is no longer dependent directly on the initial state: $p_f(\lambda|T)$ is* only an average over the true state-to-state cross-sections σ_{if}:

$$p_f(\lambda|T) = \sum_i b_i(T)\sigma_{if}(\lambda) \qquad (5)$$

This equation is used in Section 7 to synthesize temperature-dependent product-state distributions from real state-to-state cross-sections.

In most photodissociation experiments considerable averaging over initial rotational states is involved. In the case of H_2O at room temperature approximately 100 different quantum states contribute to the product-state distributions. Even in a jet-cooled probe of 50 K five to ten states still remain in the beam and will contribute to averaging. A preparation of single rotational states by jet cooling is obtained only in a few cases.[29] The averaging is even more dramatic for larger or heavier molecules with smaller rotational spacings.

The total-absorption cross-section contains even more averaging, *i.e.* summation over all final product states as well. This quantity is quite easy to measure experimentally and, for most stable molecules, has long been determined.[1] In terms of state-to-state cross-sections it is given by:

$$\sigma_{tot}(\lambda|T) = \sum_i \sum_f b_i(T)\sigma_{if}(\lambda) \qquad (6)$$

To determine the total-absorption cross-section theoretically, state-to-state cross-sections first have to be calculated and then averaged. Because total-absorption cross-sections are typically known as a function of wavelength, they contain important information about the excited-state potential, despite the fact that they are highly averaged quantities. Product-state distributions are often determined for a single wavelength or for only a few

[29] D. R. Miller in 'Atomic and Molecular Beam Methods', ed. G. Scoles, Oxford, New York, 1986.

particular wavelengths. They contain different information about the excited state, for example the anisotropy of the corresponding potential surface.

Experimental State Preparation

State-to-state experiments require the preparation of parent molecules in single quantum states. In cases of predissociation from well defined rotational states it is possible to select single rotational states simply by tuning the laser to the appropriate frequency.[30] For direct dissociation, however, rotational states have to be selected in the ground electronic state, which is a more difficult problem and requires other experimental methods.

The basic trick of the state-to-state experiment under consideration is to study the photodissociation of vibrationally excited states. The population of vibrationally excited states is negligible at room temperature, at least in the case of H_2O. Preparation of single rotational levels in vibrationally excited states is possible by various methods, including direct i.r. excitation,[21] overtone pumping,[31] stimulated emission pumping,[32] and, probably even more promising, coherent anti-Stokes Raman scattering (CARS).[33] This leads to a situation in which only a single rotational level is populated in a vibrationally excited state. If we now manage to dissociate only vibrationally excited species, *products will be exclusively formed from this single rotational state.*

The trick of dissociating only vibrationally excited molecules is very simple: the photolysis wavelength is chosen to be in the leading edge of the absorption band, where absorption of vibrational ground-state molecules is much smaller than absorption of vibrationally excited species. The method works only if the competing product formation from the vibrational ground state is small. If many more products (by several orders of magnitude) are formed from the ground state, the noise in the ground-state signal will exceed the signal from vibrationally excited states and the experiment becomes impossible.

In most cases of photodissociation in the first absorption band only a small fraction of the molecules is vibrationally excited and the rest of the population is still in the vibrational ground state. In the case of water H_2O is prepared in single rotational levels of the asymmetric stretch mode with a tunable i.r. laser around 2.7 μm. The excitation efficiency for the asymmetric stretch mode is estimated to be about 5%, and a fraction of about 5% of the molecules is assumed to be in a particular (lower) rotational state.[22] This leads to a situation in which *only a fraction of* 2.5×10^{-3} *of all* H_2O *molecules is vibrationally excited.* In the experiment we form approximately the same amount of OH products from the few vibrationally excited

[30] A. Hodgson, J. P. Simons, M. N. R. Ashfold, J. M. Bayley, and R. N. Dixon, *Mol. Phys.*, 1985, **54**, 351; M. N. R. Ashfold, J. M. Bayley, and R. N. Dixon, *Chem. Phys.*, 1984, **84**, 35.
[31] R. A. Copeland and D. R. Crosley, *J. Chem. Phys.*, 1984, **81**, 6400.
[32] D. E. Reisner, P. H. Vaccaro, C. Kittrell, R. W. Field, J. L. Kinsey, and H. L. Dai, *J. Chem. Phys.*, 1982, **77**, 573.
[33] D. A. King, R. Haines, N. R. Isenor, and B. J. Orr, *Opt. Lett.*, 1983, **8**, 629.

molecules as from the many molecules in the ground state, *i.e.* in the LIF experiment the signal for direct 193 nm dissociation is of the same order of magnitude as the signal induced by the i.r. laser. This implies that the absorption is approximately 400 times stronger for the asymmetric stretch than for the ground state. Obviously a large enhancement is required to make the experiment work. The theoretically calculated absorption cross-sections predict an enhancement of 450 for the excitation to the asymmetric stretch at $\lambda = 193$ nm.[34] This is in good qualitative agreement with the above experimental estimates.

In principle we study a different process with this method, *i.e.* the photodissociation of *vibrationally excited* molecules. However, the fragmentation still proceeds along the same excited-state potential surface; the vibrational excitation prior to dissociation changes essentially initial conditions. In the case in question the rotational and electronic fine-structure distributions do not depend on the initial vibrational-stretch state.

The advantages of this method of state preparation compared to jet cooling are, first, that only a single quantum state is prepared and, second, that the quantum state can be selected. Jet cooling allows only the preparation of ground-state molecules whereas this method additionally allows the preparation of higher rotational states.

3   Final-State Interaction and the Franck–Condon Limit

In this section the relationship between the validity of the Franck–Condon limit and the strength of final-state interaction is discussed in general terms, and it will be seen that this is closely related to the memory of the initial state.

The Two Different Steps

That the strength of final-state interaction is closely related to the validity of the FC limit may be clarified by a closer look at equation (2). *The photodissociation process consists of two very different steps.* In the first step a photon is absorbed and an electron is excited to a higher-lying orbital. This is a sudden process in which the configuration of the nuclei remains unchanged. In the second step the nuclei begin to move and, as in any other movement of heavy nuclei, couplings between the various degrees of freedom lead to a more or less strong redistribution of quantum states. This effect is called final-state interaction. Its strength is governed by the excited-state potential and may be different for different degrees of freedom. *According to the strength of final-state interaction two limiting cases are obtained in which either step 1 or step 2 dominates the product-state distributions.*

If final-state interaction is extremely weak the product-state distribution is governed exclusively by the first step, simply because a negligible coupling cannot lead to a redistribution of quantum states. It will be seen that in this

[34] V. Engel, R. Schinke, and V. Staemmler, to be published.

case a memory of the initially prepared quantum state in the parent molecule can be found in the product-state distribution.

On the other hand, a strong final-state interaction can cause an almost complete redistribution of those states that are prepared in the first step. In this case the second step dominates and the memory of the initially prepared state can be lost. The limiting case of zero final-state interaction, in which the product-state distributions are exclusively determined by the first step, is called the FC limit. For example, the photodissociation of H_2O in the first absorption band belongs to the first case for rotation and electronic fine structure and to the second case for vibration. This demonstrates that the memory of the initial state does indeed depend upon the particular degree of freedom.

Model for Pure Rotational Excitation

A few basic principles are discussed below to illustrate the origin of the validity of the FC limit for the case of pure rotational excitation. Because there is extensive literature on the theory of photodissociation dynamics,[28,35,36] only aspects that are relevant to the case of H_2O are examined here.

To describe the photodissociation process, the usual Jacobi co-ordinates that are appropriate for scattering calculations (see Figure 1) are employed: R, the vector from A to the centre of mass of BC, and r, the intermolecular BC vector. The ground (bound) state can also be described in these co-ordinates.[37,38] For simplicity we assume BC to be a rigid rotor ($r = r_e$ throughout the collision) and total angular momentum to be zero in both electronic states. The problem is then reduced to two co-ordinates: R and the orientation angle γ, the angle between R and r. In addition, the transition dipole function is assumed to be constant.

Classical Theory. The application of classical theory is straightforward. Because the excitation step 1 [equation (2)] is very sudden, the classical co-ordinates and momenta remain fixed. In the second step the particles are on the strong, repulsive, excited-state potential surface and immediately start to repel each other. The motion on this potential-energy surface, $V_{ex}(R,\gamma)$, is described in classical mechanics by the Hamilton function:[39]

$$\hat{H}(R,\gamma,P,j) = \frac{P^2}{2\mu} + Bj^2 + \frac{j^2}{2\mu R^2} + V_{ex}(R,\gamma) \tag{7}$$

where P and j are the momenta conjugate to R and γ, respectively, B is the

[35] K. F. Freed and Y. B. Band in 'Excited States', ed. E. C. Lim, Academic Press, New York, 1979, Vol. 3, p. 109.
[36] G. G. Balint-Kurti and M. Shapiro in 'Photodissociation and Photoionization', ed. K. P. Lawley, Wiley, New York, 1985, p. 403.
[37] M. Shapiro and G. G. Balint-Kurti, *J. Chem. Phys.*, 1979, **71**, 1461.
[38] G. G. Balint-Kurti and M. Shapiro, *Chem. Phys.*, 1981, **61**, 137; *Chem. Phys.*, 1982, **72**, 456.
[39] C. W. McCurdy and W. H. Miller, *J. Chem. Phys.*, 1977, **67**, 463.

rotational constant of the rotor, and μ is the reduced mass of the A–BC system. The corresponding equations of motion are:

$$\frac{dR}{dt} = \frac{P}{\mu} \qquad\qquad \frac{d\gamma}{dt} = 2j\left(B + \frac{1}{2\mu R^2}\right)$$

$$\frac{dP}{dt} = -\frac{\partial V_{ex}}{\partial R} + \frac{j^2}{\mu R^3} \qquad \frac{dj}{dt} = -\frac{\partial V_{ex}}{\partial \gamma} \tag{8}$$

The use of action-angle variables for the rotor degree of freedom is especially illustrative because *the time evolution of the molecular angular momentum j(t) is directly related to the quantity that represents the strength of final-state interaction*, i.e. *the 'torque' or 'anisotropy' of the excited-state potential $\partial V_{ex}/\partial \gamma$.*

If the final-state interaction $\partial V_{ex}/\partial \gamma$ is exactly zero, the molecular angular momentum is a constant of motion and the rotational distribution prepared in the first step is not changed in the second step. In this FC limit the product-state distribution is completely determined by the first step. A strong torque will result in large rotational excitation. The initial rotation from the first step may be small compared to the large rotation generated by the strong torque in the second step, leading to a loss of memory of the initial state. The situation is more complicated if the anisotropy is weak but non-zero. Then both the initial distribution from the first step *and* the dynamics in the second step affect the final rotational-state distribution. *Obviously the validity of the FC approximation is closely related to the strength of final-state interaction, represented for the rotational degree of freedom by the angular anisotropy of the potential.*

The anisotropy is essentially important in the angular range of the nuclear wavefunction in the ground state. If the parent molecule is initially in its ground bending state the distribution of angles γ is a narrow, Gaussian-type function centred at the equilibrium angle, γ_e, in the electronic ground state. Therefore, only a narrow range of initial orientation angles is probed and *only the anisotropy of V_{ex} around γ_e is important.* It will be seen later that this point is particularly crucial for H_2O.

Classical theory provides a very clear explanation and, in general, is simple to apply, especially in the strong coupling limit. However, in the FC limit the classical picture may not be appropriate. It will be seen that the product-state distributions are very sensitive to details of the quantum mechanical wavefunction of H_2O in the electronic ground state. The highly irregular product-state distributions found in the photodissociation of water in the first absorption band can only be explained by quantum mechanics.

Quantum Mechanical Theory. In quantum mechanics the absorption cross-section is given by Fermi's Golden Rule, which comes from a first-order perturbation treatment for the light–matter interaction.[28] The three dimensional quantum mechanical theory of photodissociation has been rigorously formulated by, for example, Balint-Kurti and Shapiro.[38] Here only a brief outline specific to the model described above is given. In particular, the total angular momenta in both electronic states are assumed to be zero, which drastically simplifies the formulation. Although this is not necessary, it is done to reduce the complexity of the problem and to demonstrate only the important principles.

The cross-section from initial rotational state i in the parent molecule to the final rotational state j in the product molecule is given by

$$\sigma_{if} \sim \nu \, | < \Psi_{gr}^i(R,\gamma) \, | \, \Psi_{ex}^j(R,\gamma) > |^2 \tag{9}$$

where $\Psi_{ex}^j(R,\gamma)$ are the dissociative wavefunctions in the excited state. The functions Ψ_{ex}^j are degenerate solutions of Schrödinger's equation for the same energy but distinguished by the asymptotic behaviour as $R \to \infty$:

$$\Psi_{ex}^j(R,\gamma) \underset{R \to \infty}{\sim} k_j^{-\frac{1}{2}} e^{ikjR} Y_{j0}(\gamma,0) + \sum_{j'} A_{jj'} e^{-ikj'R} Y_{j'0}(\gamma,0) \tag{10}$$

The wavenumbers are defined by $k_j^2 = 2\mu[E - Bj(j + 1)]$ and $E = E_i + h\nu$. The Y_{j0} are spherical harmonics and represent the rotor states. Only in-plane motion is considered ($m_j = 0$). Expanding the total wavefunctions Ψ_{ex}^j as:

$$\Psi_{ex}^j(R,\gamma) = R^{-1} \sum_j \chi_{jj'}(R) Y_{j'0}(\gamma,0) \tag{11}$$

and inserting into Schrödinger's equation yields the usual set of coupled equations:

$$[-\frac{d^2}{dR^2} - k_j^2 + \frac{j(j + 1)}{R^2}]\chi_{jj'}(R) + 2\mu \sum_{j'} V_{jj'}(R)\chi_{jj'}(R) = 0 \tag{12}$$

where the coupling matrix elements are defined by:

$$V_{jj'}(R) = 2\pi \int_0^\pi d\gamma \sin\gamma \, Y_{j0}(\gamma,0) V_{ex}(R,\gamma) Y_{j'0}(\gamma,0) \tag{13}$$

It is assumed that the wavefunction in the ground state can be written as:

$$\Psi_{gr}^i(R,\gamma) = \Psi_{gr}^i(R)\Psi_{gr}^i(\gamma) \tag{14}$$

The angular part of the ground-state wavefunction is then expanded in the basis of free rotor states:

$$\Psi_{gr}^i(\gamma) = \sum_{j''} a_{j''}^i Y_{j''0}(\gamma,0) \tag{15}$$

with:

$$a_{j''}^i = <\Psi_{gr}^i \, | \, Y_{j''0}> = 2\pi \int_0^\pi d\gamma \sin\gamma \, \Psi_{gr}^i(\gamma) Y_{j''0}(\gamma,0) \tag{16}$$

Inserting equations (15), (14), and (11) into equation (9) yields for the cross-section the simple expression:[38]

$$\sigma_{ij} \sim v \, | \, \sum_{j'} a_{j'}^i \int_0^\infty dR \chi_{jj'}^*(R) \Psi_{gr}^i(R) \, |^2 \tag{17}$$

Equation (16) may be interpreted in the following, certainly somewhat oversimplified, way: $|a_{j'}^i|^2$ is the probability for the preparation of state j' in the excited state and results exclusively from the first step. The integral is related to the probability of obtaining the final state j from the initially prepared state j' in the second scattering step. To determine cross-sections, the radial functions $\chi_{jj'}(R)$ have to be calculated, *i.e.* the full scattering problem on the excited-state potential surface has to be solved.

As discussed above, the limit of weak final-state interaction is obtained in classical mechanics if the anisotropy $\partial V_{ex}/\partial \gamma$ is zero. This implies that V_{ex} does not depend on γ, and in this case the matrix elements in equation (13) become diagonal $[V_{jj'}(R) = V_{ex}(R)\delta_{jj'}]$: the off-diagonal potential matrix elements become zero if the classical final-state interaction becomes zero. This illustrates that *the magnitude of the off-diagonal potential matrix elements represents the strength of final-state interaction in quantum mechanics.*

If the off-diagonal potential matrix elements are zero, the set of equations (12) decouples completely, *i.e.* the different components of the total wave-functions are no longer coupled through the potential. Then the radial expansion functions are also diagonal in j and j' and the sum in the cross-section expression (17) vanishes:

$$\sigma_{ij} \sim v \, | \, a_j^i \int_0^\infty dR \chi_{jj}^*(R) \Psi_{gr}^i(R) \, |^2 \tag{18}$$

This situation corresponds to elastic scattering in the exit channel. Because there is no redistribution among rotational states in the second step, the product-state distribution is determined already in the first step. This is the FC limit in quantum mechanics.[35,40]

If V_{ex} is anisotropic, *i.e.* if the coupling elements are appreciably different from zero, the full set of coupled equations must be solved numerically and a simple interpretation is generally not possible. Qualitatively, the distortion of the initial FC distribution increases with increasing anisotropy of V_{ex}. In the strong coupling limit, when the final distribution of internal states is exclusively determined by the dynamics in the exit channel, the exact close-coupling calculations become difficult to interpret and classical theory may be more explanatory.

This FC limit can be further simplified. In equation (18) the problem of elastic scattering on the excited-state potential surface still needs to be solved, *i.e.* the radial elastic-scattering wavefunctions have to be determined.

[40] M. D. Morse, K. F. Freed, and Y. B. Band, *J. Chem. Phys.*, 1979, **70**, 3604; M. D. Morse and K. F. Freed, *Chem. Phys. Lett.*, 1980, **74**, 49; *J. Chem. Phys.*, 1981, **74**, 4395; *J. Chem. Phys.*, 1983, **78**, 6045.

If the energetic difference between the various final j states is negligible (as is usually the case for rotations), the wavenumbers k_j and therefore also the radial wavefunctions and the integrals in equation (18) are independent of j. Then the distribution of final rotational states is simply proportional to:

$$\sigma_{ij} \sim |a_j^i|^2 \tag{19}$$

i.e. it is determined by the *expansion of the ground-state wavefunction in terms of the free rotor eigenstates*, according to equation (15). If final-state interaction is negligible, the product-state distribution is identical to that prepared in the first step. In the FC limit the final-state distribution is a direct 'reflection' of the parent nuclear wavefunction,[40,41] which is not 'destroyed' in the second step.

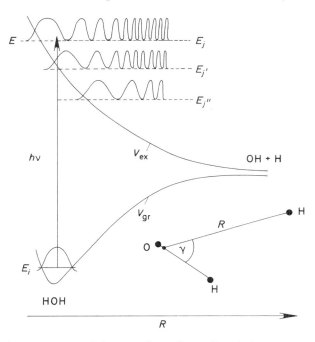

Figure 1 *Schematic potential diagram for a direct dissociation process with ground- and excited-state potential surfaces, V_{gr} and V_{ex}. The radial nuclear wave-function in the ground state, $\Psi_{gr}(R)$, and several components, $\chi_{jj}(R)$, of the nuclear wavefunction in the excited state are shown. The E_j are the channel energies $E - Bj(j + 1)$, greatly enlarged here. The co-ordinate system for the triatomic model including only rotational excitation is shown in the inset*

If the wavenumbers for the various internal states are significantly different (as is usually the case for vibrations and as is indicated in Figure 1), the radial integral in equation (18) strongly modifies the 'pure' FC distribution from equation (19).

[41] M. Shapiro, *Chem. Phys. Lett.*, 1981, **81**, 521; M. S. Child and M. Shapiro, *Mol. Phys.*, 1983, **48**, 111.

The FC Limit in the General Case

In the above section the FC limit was formulated and derived for a very special case, *i.e.* only one internal degree of freedom was considered and the total angular momentum in the electronic ground state was assumed to be zero. In this case it could be rigorously shown that the state-to-state cross-section is simply proportional to the square of the coefficients a_j^i. These coefficients describe the angular part (*i.e.* the bending wavefunction) in terms of the rotor eigenfunctions Y_{j0}.

The situation is certainly more complicated if initial rotational excitation and/or more degrees of freedom are involved. However, even in such cases the general idea of the FC theory is still valid. First, the initial-state wavefunction $|i>$ must be expanded in terms of product-state wavefunctions $|f>$ according to:

$$|i> = \sum_f |f> <f|i> \tag{20}$$

which is always possible because the $|f>$ form a complete set (if all degrees of freedom are included). Then the state-to-state cross-sections are simply given by:

$$\sigma_{if} \sim |<i|f>|^2 \tag{21}$$

i.e. the square of the expansion coefficients is the state-to-state photodissociation cross-section. Equations (20) and (21) are obvious extensions of equations (15) and (19).

It may be illustrative to discuss the simplicity of the FC limit. In the expansion of a low rotational state of H_2O in terms of the OH product eigenfunctions according to (20), only a few rotational states of OH contribute significantly. More and higher rotational states of the OH product are necessary to describe higher rotational states of H_2O. This implies that the OH will be formed in low (or high) rotational states if 'cold' (or 'warm') H_2O is photolysed.

Because the wavefunction for the initial state $|i>$ of the parent molecule appears explicitly in (21), the state-to-state dissociation cross-sections depend directly upon the initial state. *In the FC limit a strong memory of the initial state is expected.* As demonstrated below, the state-to-state cross-sections are very sensitive to details of the nuclear wavefunction in the ground state.

The FC Limit Including Electronic Fine Structure

The preceding Franck–Condon theories included only vibrational and rotational degrees of freedom. Although the FC limit has also been applied to the case of H_2O,[42] a reasonable comparison between experiment and theory was not possible, simply because the electronic fine structure was

[42] J. A. Beswick and W. M. Gelbart, *J. Phys. Chem.*, 1980, **84**, 3148.

neglected. In the experiment the distributions for the two Λ-doublet states were found to be different and it was by no means clear which distribution should be compared with the FC result. The only reasonable approach would be first to average the experimental results over the electronic fine structure. However, this gives poor agreement between theory and experiment.[22]

The explicit inclusion of the electronic fine structure in the FC approximation has been worked out recently by Balint-Kurti.[26] The electronic structure of both the parent and the products has been included in the treatment, *i.e.* OH wavefunctions[13] intermediate between Hund's cases *a* and *b* have been used along with the spin of the hydrogen atom. The theory leads to a rather complicated-looking expression for the state-to-state cross-sections, which is not repeated here but given in the original references.[26,27] This expression is analytical and contains many angular momentum coupling elements as well as the expansion coefficients of the nuclear wavefunction of water in the ground state. The latter must be determined numerically.[21] Since H_2O is an asymmetric top molecule the calculation of the bound-state wavefunction is not insignificant, especially for higher rotational states. Although the expression looks complicated, the basic idea of the theory is as simple as outlined above. The complications arise only from the coupling of many angular momenta.

Unfortunately, the theory as it stands so far holds only in such cases as H_2O or H_2S, where the fragmentation evolves along an A'' surface to a diatomic molecule in a $^2\Pi$ state and an atom in a ^2S state. It may already be wrong in rather similar cases like the photodissociation of NO_2, where the diatomic (NO) is still formed in a $^2\Pi$ state, the atom being formed in different spin states [$O(^3P)$[43] or $O(^1D)$[44]]. To understand the selective population of electronic fine-structure states in other cases, an extension of the present type of FC theory is urgently required. The important clue in the above theory is the prediction of electronic fine-structure distributions, which until now could not be obtained by any other method. The inclusion of electronic structure in a full numerical calculation is possible in principle[26,45] but prohibited by the extremely large numerical effort.

Probably the most interesting aspect of the present FC theory is that it predicts *selective* population of electronic fine-structure states, as has been found in many photodissociation studies in recent years, for Λ-doublet states[46] or for spin states.[47] An important question is whether this selectivity

43 U. Robra, thesis, Universität Bielefeld, West Germany, 1984; K. H. Welge, *NATO ASI Ser., Ser. B.* 1984, **105**, 123.

44 L. Bigio and E. R. Grant, *J. Phys. Chem.*, 1985, **89**, 5855.

45 M. H. Alexander and S. L. Davis, *J. Chem. Phys.*, 1983, **79**, 227; M. Alexander, *J. Chem. Phys.*, 1982, **76**, 5974.

46 R. Mariella, B. Lantzsch, V. T. Maxson, and A. C. Luntz, *J. Chem. Phys.*, 1978, **69**, 5411; K. Kleinermanns and J. Wolfrum, *J. Chem. Phys.*, 1984, **80**, 1446; R. Vasudev, R. N. Zare, and R. N. Dixon, *J. Chem. Phys.*, 1984, **80**, 4863.

47 I. Nadler, H. Reisler, and C. Wittig, *Chem. Phys. Lett.*, 1984, **103**, 451; F. Shokoohi, S. Hay, and C. Wittig, *Chem. Phys. Lett.*, 1984, **110**, 1; I. Nadler, D. Mahgerefteh, H. Reisler, and C. Wittig, *J. Chem. Phys.*, 1985, **82**, 3885.

originates from the first or from the second step of the dissociation process. *A selective population of electronic fine-structure levels may be expected in many cases from the first step already.*

The FC theory is always a first guess for the product-state distributions. It will actually predict the correct distribution if the final-state interaction is negligibly small. Although the selectivity generated in the first step can be washed out by the final-state interaction in the second step, this need not necessarily be the case. As an example, the high rotational states found in the photodissociation of MeONO[48] (see Chapter 5) clearly indicate a strong final-state interaction for rotation; nevertheless, a pronounced Λ-doublet preference is found. This preference clearly originates from the first excitation step.

It may be interesting to look at this problem from a more general point of view. The state-to-state cross-sections in the FC theory are the expansion coefficients of the parent molecule in terms of product-state eigenfunctions. This expansion can be very complicated, because the wavefunction of the parent molecule can be very different from the product-state eigenfunctions.

This holds in particular for the electronic structure, which is often very different for parent molecules and products. For example, we consider the dissociation of water in the first absorption band, where H_2O fragments into OH and H. In H_2O the spins are either parallel or antiparallel, leading to singlet or triplet states. The electronic wavefunctions are either symmetric or antisymmetric, leading to the very different potential surfaces A' and A''.[49] Because the coupling among the electrons is much stronger than the coupling of the electrons to the nuclear degrees of freedom, the Born–Oppenheimer approximation is valid. In the products, OH and H, the situation is very different. After the break-up of the excited complex, unpaired electrons remain on the OH and on the H atom. For example, the spin of the H atom is now uncoupled from the spin of OH. The electronic angular momentum in OH couples to the nuclear rotation and the resulting wavefunctions are rather complicated.[13] These very different wavefunctions are then used to represent the smooth nuclear wavefunction of the parent molecule! It seems obvious that this expansion produces structured distributions, and therefore some selectivity for electronic fine-structure degrees is not surprising at all.

General Features of the FC Limit

To conclude the general part of this review, the most important features that are expected in the two limiting cases of direct dissociation processes, *i.e.* weak and strong coupling limit, are summarized briefly. These two limits yield very characteristic product-state distributions, and an attempt is made

[48] O. Benoist d'Azy, F. Lahmani, C. Lardeux, and D. Solgadi, *Chem. Phys.*, 1985, **94**, 247 and references therein.

[49] G. Herzberg, 'Molecular Spectra and Molecular Structure', van Nostrand, New York, 1950, Vols. 1 and 2.

to give a few qualitative rules that might be helpful in understanding experimental results. A rigorous classification of coupling cases is certainly best obtained on the basis of dynamical calculations using known potential-energy surfaces. However, if experimental product-state distributions agree quantitatively with the appropriate FC prediction, this indicates directly that the final-state interaction is weak. The discussion will be restricted to rotational and electronic degrees of freedom.

In the FC limit the following features are expected:

(a) According to equations (15) and (21) the state-to-state cross-section is a direct 'reflection' of the bound wavefunction of the parent molecule. Since usually few product-state wavefunctions are needed to describe the initial state, the resulting product distributions are very often 'cold', *i.e.* only the lowest states are populated. The amount of internal product excitation gradually increases for initially excited states of the parent molecule.[40,42]

(b) The product-state distributions depend very strongly upon the initial state of the parent molecule. This will result, for example, in product-state distributions that depend on temperature.

(c) In the simplified FC limit the product-state distributions do not depend at all upon the photon energy. In the modified FC limit a slight wavelength dependence is predicted. A strong wavelength dependence is only possible for strong final-state interactions.

In the case of strong final-state interaction the rotational-state distributions will peak at high j, indicating a strong classical torque. *These distributions can be qualitatively described by the rotational-reflection principle.*[50] In contrast to the FC limit, the rotational-reflection principle predicts that the rotational distributions will vary with the photolysis wavelength. On the other hand, a large torque in the second step will considerably wash out details of the preparation in the first step and the final-state distribution may no longer depend on the initial state. The memory of the initial state, which is so pronounced in the FC limit, is more or less lost.

It is interesting to take a closer look at the temperature-dependent product-state distributions $p_f(\lambda|T)$ in equation (5), because *a comparison of product-state distributions for different temperatures can already reveal whether state-to-state photodissociation cross-sections depend upon the initial state or not.* To understand this, consider two different cases. In the first case the state-to-state cross-sections are assumed to be independent of the initial state, *i.e.* there is no memory of the initial state. This can be true for strong final-state interaction, as discussed above. In this case the product-state distribution will not depend on temperature, simply because σ_{if} can be extracted from the sum in equation (5), and the remaining sum over the Boltzmann weights is normalized to unity, independent of temperature. In the second case the state-to-state cross-sections are assumed to depend on

[50] R. Schinke, *J. Chem. Phys.*, 1986, **85**, 5049; R. Schinke and V. Engel, *Faraday Discuss. Chem. Soc.*, 1986, **82**, in press.

the initial state, as for example for weak final-state interaction. Then the product-state distributions can depend on temperature, although the temperature dependence may be washed out by averaging. *If the product-state distributions are found experimentally to be temperature dependent, the conclusion is that the state-to-state cross-sections depend upon the initial quantum state.* Although not strictly conclusive, a temperature independence of an experimental product-state distribution suggests strongly that the state-to-state cross-sections do not depend on the initial state.

There are several examples that fit qualitatively into this scheme of weak and strong final-state interaction. Many rotational-state distributions are found to be quite narrow and highly inverted (for a selection see, for example, Figure 1 of ref. 50). In these cases final-state interaction is obviously strong. In some of these cases (ICN,[47] MeONO[48]) experiments have been carried out for both room-temperature and jet-cooled species. As expected from the above arguments, no difference was found. In some instances the experiments were conducted at different photolysis wavelengths and the distributions were found to be shifted to higher states with increasing wavelength, at least in qualitative agreement with the rotational-reflection principle.[50]

In several other cases the rotational-energy content in the products was found to be small. For H_2O and H_2S[51] the photolysis was studied for both room-temperature and jet-cooled species. The large difference in the temperature-dependent product-state distributions observed for these examples indicates that the state-to-state cross-sections depend on the initial state of the parent molecule. This strong dependence of the state-to-state cross-sections on the initial state is impressively demonstrated below for the photodissociation of H_2O. The photolysis wavelength turns out to be rather unimportant, although the product-state distributions do change slightly with wavelength. A small wavelength dependence of the rotational-state distributions can be explained, if the FC theory of equation (18) is used instead of the simplified FC limit in equation (19). Energetic effects are primarily expected at the steep onset of the absorption cross-section at large wavelengths.

4 The Physical Situation of the Photodissociation of H_2O in the First Absorption Band

In this section the physical situation for the photodissociation of H_2O in the first absorption band is described. The quantum states of H_2O and OH are explained and some relevant features of the *ab initio* potential surface for the excited state are discussed. It will be explained why the state-to-state cross-sections separate into a wavelength-independent rotational part and a wavelength-dependent vibrational part.

[51] W. G. Hawkins and P. L. Houston, *J. Chem. Phys.*, 1982, **76**, 729.

Correlation Diagram and Electronic Structure

In the photodissociation of water in the first absorption band a particularly clear physical situation is met. This is best explained with the correlation diagram of Figure 2, which shows the deep ground-state well of water $\tilde{X}^1A_1$ (or A') and the first electronically excited state $\tilde{A}^1B_1$ (or A''). The energetic positions of the next-higher-lying electronic states are also indicated.

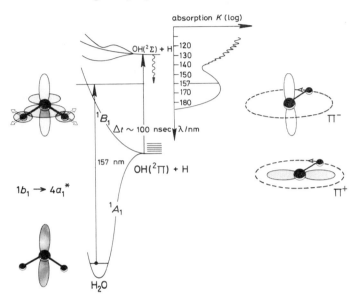

Figure 2 *Schematic correlation diagram for the lower electronic states of water. The electronic ground state $\tilde{X}^1A_1$, the strong repulsive first electronically excited state $\tilde{A}^1B_1$, and (roughly) the position of a few other higher-lying states are shown. The electronic structure of H_2O in the ground and first excited state is sketched in the lower and upper left-hand corners, respectively. The two possible orientations of the unpaired electron relative to the OH rotation plane are shown on the right-hand side. The upper right-hand corner gives qualitatively the absorption cross-section for H_2O*

(Reproduced with permission from *J. Chem. Phys.*, 1984, **80**, 2548)

Obviously, the surface $\tilde{A}^1B_1$ is well isolated from all other surfaces. It is responsible for the first absorption band that extends from ~ 130 nm up to 190 nm and is given in the upper right-hand corner of Figure 2. In this energy range there are no crossings with other singlet states,[52] precluding non-adiabatic transitions as the fragments separate. This implies in particular that the fragmentation is direct and that no predissociation occurs. The photodissociation leads exclusively to ground-state products [OH($^2\Pi$) and H(^2S)]. Although the formation of O(^1D) + H$_2$ is energetically allowed, it is forbidden by a barrier in the $\tilde{A}^1B_1$ potential surface.[10]

[52] M. B. Robin, 'Higher Excited States of Polyatomic Molecules', Academic Press, New York, 1974, Vols. 1 and 2; 1985, Vol. 3.

The electronic structure of the excited complex is well known for the photodissociation of H_2O in the first absorption band, in which the most weakly bound electron is excited to the next-higher-lying orbital. The most weakly bound electrons are in the so-called $1b_1$ orbital, essentially a pure $p\pi$ lobe perpendicular to the nuclear plane. This orbital is shown in the lower left-hand corner of Figure 2. Because one of the two electrons in this orbital is excited to the next-higher orbital, there will be *one unpaired $p\pi$ electron perpendicular to the H_2O plane in the excited complex*.

The next-higher orbital is the so-called $4a_1{}^*$ orbital, a strong antibonding orbital symmetric to the plane of the nuclei. This orbital is responsible for the strong repulsive character of the $^1B_1$ state. A sketch of the electronic structure in the excited complex is given in Figure 2 in the upper left-hand corner. It shows the unpaired $p\pi$ electron together with the $4a_1{}^*$ orbital. This shape of the excited complex is used later to explain the origin of the selective population of Λ-doublet states.

The Quantum States of H_2O and OH

To understand the quantum states that are involved in the state-to-state experiment under discussion, consider the energy-level diagrams of H_2O and OH. The H_2O molecule in the electronic ground state has a closed electronic shell. The quantum states for vibrational motion are given by a set of quantum numbers (v_1, v_2, v_3): v_1 for the symmetric stretch, v_2 for the bending motion, and v_3 for the asymmetric stretch. The quantum states for the rotation are total angular momentum J, the projection K_a of J on the a-axis, and the projection K_c of J on the c-axis. Each rotational state is classified as $J_{K_aK_c}$. Figure 3a shows the resulting energy-level diagram of H_2O rotational states in the vibrational ground state $(0,0,0)$.

In the photodissociation of H_2O in the first absorption band the product OH is formed in its $^2\Pi$ electronic ground state. The open electronic shell leads, in addition to rotation and vibration, to electronic fine-structure degrees of freedom. The configuration for the outer electrons of OH in the ground state $^2\Pi$ is ... $(2p\sigma)^2(2p\pi)^3$, which implies that three electrons are in the two $p\pi$ orbitals perpendicular to the OH internuclear axis. In the following we will discuss this situation as if there were only one unpaired electron in one $p\pi$ orbital. Although this simplification has caused some controversy,[53] all the important conclusions remain unchanged. The unpaired electron in the $p\pi$ orbital determines the open-shell character of the OH molecule. This unpaired $p\pi$ electron possesses both electronic orbital angular momentum $(\Lambda = 1)$ and spin $(S = \frac{1}{2})$. Depending on the coupling of Λ and S, two multiplet states $^2\Pi_{\frac{1}{2}}$ and $^2\Pi_{\frac{3}{2}}$ are obtained for the non-rotating molecule.

In addition to the splitting into two multiplet states, there is another splitting that is caused by the unpaired electron. The origin of this splitting is

[53] M. H. Alexander and P. J. Dagdigian, *J. Chem. Phys.*, 1984, **80**, 4325.

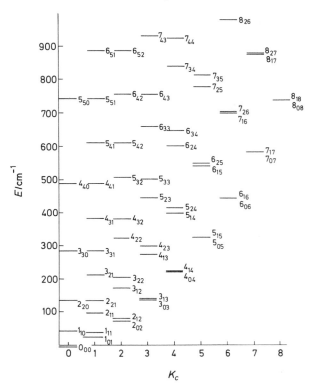

Figure 3a *The energy-level diagram of* H_2O. *The rotational states are classified with* $J_{K_aK_c}$ *and organized with increasing* K_c

explained on the right-hand side of Figure 2, where the unpaired electron is shown for the case of a rotating OH molecule. The unpaired electron in OH can be oriented in different ways relative to the rotation plane: it can be either perpendicular to the rotation plane or in the rotation plane. Either way, the lobe has to be perpendicular to the internuclear axis. These two possible orientations represent the important difference between the Λ-doublet states of OH.

If the lobe is perpendicular, the *electronic* wavefunction of the OH molecule will be antisymmetric with respect to a reflection at the rotation plane. If the lobe is in plane, the electronic wavefunction will be symmetric. This different reflection symmetry yields two states of different electronic symmetry, the so called Λ-doublet states. Although these states are almost degenerate, they are of particular interest for the product-state distributions, as will be seen later on. The reader should keep the orientation of the unpaired $p\pi$ electron in mind, because it is responsible for the selective formation of OH in Λ-doublet states.

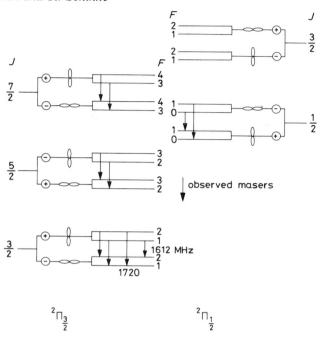

Figure 3b *The energy-level diagram of* OH. *It shows for the* $^2\Pi_{\frac{3}{2}}$ *and the* $^2\Pi_{\frac{1}{2}}$ *multiplet states the first few rotational states. Each rotational state is split into two Λ-doublet states. Parity and orientation of* $p\pi$ *lobe are indicated for each state by the* $+/-$ *and the symbols* ∞ *and* $\S$, *respectively. The Λ-doublet states split again into hyperfine levels. The arrows indicate the transitions at which astronomical* OH *masers are observed*

With this information the level diagram of OH in Figure 3b may be easily understood. It shows for the two multiplet states $^2\Pi_1$ and $^2\Pi_3$ a series of rotational states, each of which is split again by Λ-doubling. Due to the nuclear spin of the hydrogen atom a further hyperfine splitting occurs, which is also shown in Figure 3b. For each of the levels in Figure 3b the orientation of the unpaired $p\pi$ orbital is given by the symbols $\S$ and ∞. The symbol $\S$ indicates an orientation perpendicular to the rotation plane and the symbol ∞ indicates an orientation in the rotation plane. These symbols are only used to classify the more symmetric or asymmetric character of the electronic wavefunction. The true electronic densities look different.

The orientation given here is exactly opposite to the assumptions of all earlier OH maser theories, where it was assumed that the lobe is in plane for the upper Λ-doublet in $^2\Pi_3$. Here, the $p\pi$ lobe is always perpendicular to the OH rotation plane for the upper Λ-doublet in $^2\Pi_3$. In the other multiplet state, $^2\Pi_1$, the reverse is true: the lobe is perpendicular to the plane for the lower Λ-doublet and in the plane for the upper Λ-doublet (for the first five rotational states). The consequences of this erroneous assumption will be discussed later on.

Dissociation of Water in the First Absorption Band

The Excited-State Potential Surface

In the following, characteristic features of the excited-state potential-energy surface that are of central importance for both the qualitative and the quantitative understanding of the experimental results are discussed.

Staemmler and Palma[10] calculated the potential-energy surface of the $\tilde{A}^1 B_1$ state with quantum mechanical *ab initio* methods including electron correlation. More than 200 geometries in a wide range of co-ordinate space were sampled to allow full three-dimensional dynamical calculations including all nuclear degrees of freedom. The two OH distances, R_{OH}, were varied between 1.6 a_0 and 4.0 a_0 and the HOH bending angle α was varied between $0°$ and $180°$. The *ab initio* points were fitted to an accurate analytical expression.[34]

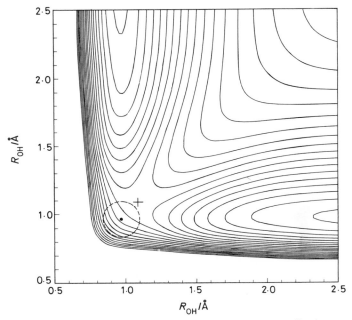

Figure 4 *Contour diagram of the excited-state potential surface $\tilde{A}^1 B_1$ for the HOH bending angle $\alpha = 104.5°$ plotted against the two OH distances. The dashed circle indicates the width of the ground-state wavefunction; the cross marks the position of the barrier*

Figure 4 shows a contour diagram of this potential-energy surface for $\alpha = 104°$, *i.e.* the equilibrium angle of the ground electronic state. The energy normalization is such that H + H + O corresponds to zero. The energy of the H + OH (r_e) asymptote is then $E_\infty = -4.621$ eV, and the energy of the (000|J = 0) level in the $\tilde{X}$ state is -9.500 eV as calculated from the empirical potential of Sorbie and Murrell.[54] The potential is symmetric with respect to the two OH distances. It has a barrier of 1.98 eV (with respect to E_∞) at

[54] K. S. Sorbie and J. N. Murrell, *Mol. Phys.*, 1975, **29**, 1387; *Mol. Phys.*, 1976, **31**, 905.

$R_{OH} = 1.09$ Å for $\alpha = 104°$. This barrier varies only slightly with the angle in the region $80° < \alpha < 120°$, which is primarily probed by dissociation following excitation from the ground state.[34] Also indicated in Figure 4 is the range of the ground vibrational wavefunction in the $\tilde{X}$ state. According to Figure 1, absorption from the ground to the excited state will occur only if the dissociative wavefunction has appreciable overlap with the narrow spatial region defined by the ground-state wavefunction. This implies that absorption is strong only at energies well above the barrier. Because the potential surface is clearly strongly repulsive along the dissociation co-ordinate, the dissociation must be fast and direct and along a single excited-state potential surface.

Another important feature of the excited-state potential surface is its preference for the same angle, $\alpha \approx 104°$, that H_2O has in its ground state. The strong torque that is known to cause large rotational excitation in many linear-bent transitions is missing here. As discussed above, this leads to very small final-state interaction and to the validity of the FC limit for the rotational degree of freedom. In the photodissociation of H_2O in the first absorption band, a non-bonding electron ($1b_1$) is excited to an antibonding orbital ($4a_1^*$). The 'conservation' of the bond angle may result from the fact that the more strongly bonding electrons still govern the energetics. A similar situation may be found in other cases as well, and will lead to the validity of the FC limit there. The preference for $\sim 104°$ is demonstrated in Figure 5, which shows the angular variation of the interaction potential $V_1(R, r, \gamma) = V_{ex}(R, r, \gamma) - V_{ex}(R = \infty, r, \gamma)$. It goes to zero for large H—OH distances. R is the separation between the departing H atom and the centre of mass of the OH fragment, r is the intermolecular OH separation (fixed at its asymptotic equilibrium value in Figure 5), and γ is the angle between R and r ($\gamma = 0°$ corresponds to the collinear H—HO configuration).

The $\tilde{A}^1B_1$ potential surface has a large *overall* anisotropy, and a H + OH $(j) \rightarrow$ H + OH (j') scattering experiment would result in strong rotational excitation.[25] However, as discussed above, in photodissociation only the anisotropy above the ground-state equilibrium, $\gamma_e \sim 104°$, is important. Figure 5 clearly shows that $V_{ex}(\tilde{A}^1B_1)$ is *locally* isotropic in this angular region, even at small distances, with the consequence that scattering effects on the rotational motion of OH during dissociation are extremely weak.[25] Classically, H_2O is initially prepared in the $\tilde{A}^1B_1$ state at angles around $104°$ with the distribution of initial angles γ approximately given by $|\Psi_{gr}(\gamma)|^2$, where $\Psi_{gr}(\gamma)$ is the angular part of the ground-state wavefunction that is also shown in Figure 5. In this angular region the torque $\partial V_{ex}/\partial \gamma$ is extremely small at all internuclear distances. This has the consequence that the OH rotational angular momentum remains almost constant as it readily follows from Hamilton's equations (8).

The important conclusion of the above discussion is that the potential surface predicts an extremely weak final-state interaction for the rotational degree of freedom. As discussed above, the rotational product-state distribution will be governed exclusively by the first step and the FC limit will be

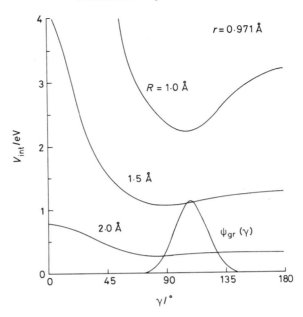

Figure 5 *Angular variation of the $\tilde{A}^1B_1$ potential-energy surface for different H—OH distances R. The other OH separation is fixed at its equilibrium value. In particular, the interaction potential $V_1(R,\gamma) = V_{ex}(R,\gamma) - V_{ex}(R = \infty)$ is shown. $\Psi_{gr}(\gamma)$ indicates the angular part of the nuclear wavefunction in the electronic ground state*

valid *for the rotational degree of freedom*. The potential surface does not predict anything about the final-state interaction for the electronic degrees of freedom (Λ-doublet and spin). However, the good agreement between experiment and the FC theory with electronic structure, which is shown later, demonstrates that final-state interaction is also small for the electronic degrees of freedom.

Separation of State-to-State Cross-sections

The photodissociation of H_2O in the first absorption band turns out to be a very favourable case for dynamical treatment. As rationalized in the preceding section, the rotational degree of freedom can be perfectly described in the FC limit. This has been tested quantitatively by comparison with dynamically coupled calculations in which, however, the electronic degrees of freedom and the two dissociation channels have been ignored.

The nuclear wavefunctions of water in the electronic ground state and for zero total angular momentum can be written approximately as:

$$\Psi_{gr}^{J_i = 0}(R, r, \gamma) = \Psi_1(R)\Psi_2(\gamma)\Psi_3(r) \tag{22}$$

where Ψ_1 and Ψ_3 describe the two OH stretch modes and Ψ_2 describes the bending motion. Equation (22) is valid for low vibrational states, and it is also valid (but more complicated) if overall rotation is included ($J_i \neq 0$). To predict rotational-state distributions within the FC limit, only the angular part of Ψ_{gr} is required, *i.e.* $\Psi_2(\gamma)$ and its expansion into proper product-state eigenfunctions, including the electronic degrees of freedom. For water in the $v_2 = 0$ bending state, $\Psi_2(\gamma)$ is a Gaussian-shaped function centred around $\gamma = 104°$.

Since $\Psi_2(\gamma)$ is, to a very good approximation, independent of the two stretch states, the state-to-state cross-sections, including the rotational quantum numbers of the parent molecule ($J_{K_aK_c}$) and the quantum numbers of the product molecule ($nN\Lambda S$), can be approximately written in the following factorized form:

$$\sigma_{nN\Lambda S}^{J_{K_aK_c}}(\lambda) = \sigma_n(\lambda)\sigma_{N\Lambda S}^{J_{K_aK_c}} \tag{23}$$

The first, wavelength-dependent, term predicts the final vibrational distribution. It may be mentioned that this term also depends strongly on the initial symmetric and asymmetric stretch states, which will be discussed elsewhere.[34] The second term predicts the final rotational and electronic fine-structure distributions. It is independent of the wavelength and the particular OH vibrational state. It depends, however, strongly upon the initial rotational state $J_{K_aK_c}$.

The rotational and electronic fine-structure part is accurately described within the FC limit, and reliable cross-sections can be predicted with moderate effort for any initial rotational state.[21] The vibrational part must be treated dynamically by rigorously solving Schrödinger's equation. These calculations are still very difficult, because both dissociation channels have to be taken into account. If rotation, vibration, and translation were strongly coupled, such calculations would be impossible to perform at the present time. However, because only low rotational states are excited in the dissociation, which is obviously a consequence of the FC limit, the so-called energy sudden approximation can be utilized. Then only vibration and translation remain coupled, and the quantitative treatment, which will be discussed in Section 6, becomes possible. *It is this situation that allows complete characterization of the photodissociation of* H_2O *in the first absorption band.*

5 Results

In this section a few experimental results for process (1) that are selected to demonstrate the characteristic features of the FC limit are discussed. It will be shown that the FC limit holds for both the rotational and the electronic fine-structure degrees of freedom. The pronounced memory of the initial state will be seen both in the temperature dependence of the product-state distributions $p_f(\lambda|T)$ and in the true state-to-state cross-sections σ_{if}. These effects will be discussed first for 'pure' rotational and then for the electronic fine-structure distributions, *i.e.* Λ-doublet and spin. In addition some model calculations demonstrate that the product-state distributions are extremely sensitive, for example to the bond angle of the parent molecule.

Rotational Distributions

For molecules with electronic fine structure the term 'rotational distribution' is not clearly defined. In this chapter it is used for the population of rotational states belonging *to the same multiplet and Λ-doublet state*. For each rotational distribution we have to specify both spin and Λ-doublet state. Another procedure would be to average over the electronic degrees of freedom and define the average as a rotational distribution.

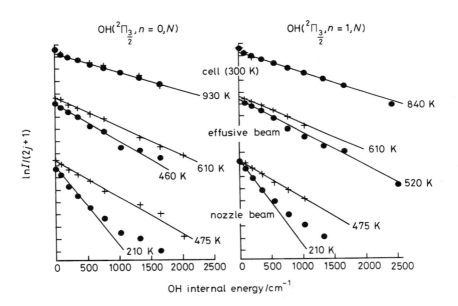

Figure 6 *Rotational distribution for the* OH *product originating from 157 nm photolysis of* H_2O *at different temperatures of the parent molecule (nozzle beam, effusive beam, and room temperature). The left-hand side is for* OH *formed in* n = 0, *the right-hand side for* OH *formed in* n = 1. *The crosses are for the upper Λ-doublet and the points for the lower Λ-doublet in* $^2\Pi_{\frac{3}{2}}$. *The lines indicate Boltzmann distributions with the temperatures given in the plot*

In this sense Figure 6 shows in a Boltzmann plot some rotational distributions for the $^2\Pi_{\frac{3}{2}}$ multiplet state, obtained in the 157 nm photolysis of the parent molecule at three different temperatures. Symbols are explained in the figure caption. These distributions represent the temperature-dependent product-state distributions of equation (5), which are averaged over state-to-state cross-sections. They are selected to demonstrate the following points:

(a) The OH rotational-state distributions are smooth and can be fitted in all cases but one (lower Λ-doublet in $^2\Pi_{\frac{3}{2}}$, jet-cooled species) by a straight line, *i.e.* they can be described by the rotational temperatures given in the plot. *The pronounced structure in the state-to-state data (see below) is already washed out by averaging over a few initial H_2O rotational states.* This point becomes even clearer in the discussion in Section 7.

(b) The OH product-state distributions do indeed depend upon the temperature of the parent molecule. This implies that the state-to-state photodissociation cross-sections depend upon the initial state of the parent molecule, as discussed above and demonstrated below.

(c) The rotational temperature in the OH product increases from ≈ 210 K to ≈ 950 K if the temperature in the parent H_2O increases from the jet-cooled species to the 300 K species. The rotational energy in the product increases obviously with the rotational energy in the parent. However, whereas the energy in the parent increases only by (less than) 300 K it increases in the OH product by 750 K. This demonstrates that the energy of the parent molecule is not simply transferred to the product.

(d) The distributions suggest a decoupling of rotational and vibrational degrees of freedom. A comparison of the data on the left-hand side for $n = 0$ and the right-hand side for $n = 1$ reveals that the rotational distributions are independent of the vibrational state. This is true for all three H_2O temperatures and supports an essential assumption of the theory.

(e) In particular for the jet-cooled H_2O, a large difference is found for the rotational distributions in the upper and lower Λ-doublet. With increasing H_2O temperature the difference becomes less pronounced. This effect indicates a selective population of Λ-doublet states, which will be discussed below.

Unfortunately the rotational temperature of the parent molecule H_2O is only well known for experiments carried out at room temperature. Because state-selective detection methods were not available for H_2O, the rotational temperature in the beams could not be measured directly. An estimate of these rotational temperatures was obtained by mixing a small amount of NO (1%) to the beams and measuring the rotational temperature of the NO component by LIF. The NO temperature was 10 K for the jet-cooled species and 260 K for the slightly cooled beam. However, the H_2O and the NO temperature can still be very different in the beam since H_2O exists in both the *ortho* and the *para* modifications.

Although the temperature-dependent product-state distributions already show some interesting features of the dissociation process, much more detailed information is contained in the true state-to-state cross-sections. The rotational-state distributions in Figures 7 and 8 originate from single rotational states of vibrationally excited H_2O as described in Section 2. These state-to-state cross-sections reveal a much more complicated picture.

Figure 7 shows experimental and theoretical rotational-state distributions for the upper and lower Λ-doublets of OH, originating from the rotational

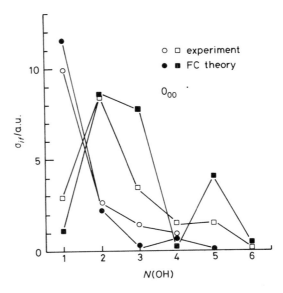

Figure 7 *Rotational distributions for the* OH *product originating from the* 193 nm *photolysis of the rotational ground state* 0_{00} *in vibrationally excited* H_2O *(0,0,1). The circles are for the* Π^+- *and the squares for the* $\Pi^- -\Lambda$-*doublet in* $^2\Pi_{\frac{3}{2}}$. *Experiment and theory are normalized to each other using the area below the* Π^- *distributions*

ground state 0_{00} of H_2O. In contrast to the temperature-dependent product-state distributions of Figure 6, these data are not smooth and cannot be described by a Boltzmann distribution. A pronounced structure, in particular for the upper Λ-doublet, shows up in both theory and experiment.

Because the rotational state 0_{00} is usually assumed to be dominant in a good nozzle beam, it is interesting to compare the distributions in Figure 7, which clearly originate exclusively from the 0_{00} state, with the distribution for the jet-cooled species in Figure 6. If the 0_{00} state really dominated the nozzle beam, the distributions should be identical. This is obviously not true and demonstrates that one has to be very careful when estimating temperatures in nozzle beams.

The experimental data in Figure 7 are compared to the predictions of the Franck–Condon theory including electronic structure.[27] The pronounced structure found experimentally is well reproduced by theory. For the lower Λ-doublet a general decrease is predicted and found experimentally. For the upper Λ-doublet the small value for $N = 1$, the increase at $N = 2$, and even the structure at $N = 5$ are similar in experiment and theory. An agreement with such pronounced structure is certainly not accidental.

Probably the most pronounced difference in the temperature-dependent product-state distributions is that *the consistent preference for the upper Λ-doublet is lost*. For $N = 1$ the lower Λ-doublet state dominates the upper Λ-

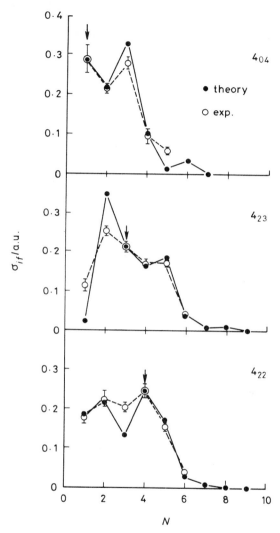

Figure 8 *Rotational distributions in the OH product originating from 193 nm photolysis of single rotational states of vibrationally excited H_2O (0,0,1); all have the same total angular momentum $J_i = 4$. The distributions are for the lower Λ-doublet in $^2\Pi_{\frac{3}{2}}$. Normalization of experiment and theory is made separately for each H_2O rotational state indicated by the arrows*

doublet by approximately a factor of 3. In almost all other cases where Λ-doublet selectivity has been observed the preference is consistently for *either* the symmetric *or* the antisymmetric Λ-doublet state. Here, and in very few other cases, the preference depends upon the final rotational state.[43,44]

Figure 8 demonstrates the strong sensitivity of the product-state distributions to the initial rotational state of the parent molecule. It shows OH product-state distributions originating from a series of single rotational states of H_2O. In this series the total angular momentum J is always the same; it is only the projection of J on the a- or c-axis that varies. The energetic difference between these states is negligible compared to the excess energy.

All of these distributions show a pronounced structure, which certainly cannot be described by a rotational temperature. The strong memory of the initial state is demonstrated by the fact that each of these distributions is different from the other. This strong memory of the initial state implies, without any theoretical assumptions, that final-state interaction must be small for rotational and electronic fine-structure degrees of freedom.

The experimental results in Figure 8 are again compared to the predictions of the FC theory.[26,27] Although there are slight discrepancies between experiment and theory, the general structures are excellently reproduced by the calculation. It should be emphasized that this comparison between experiment and theory is on a highly sophisticated level, on a state-to-state basis with six specified quantum numbers (J, K_a, K_c, Λ, N, S). We are not aware of any other comparison on such a detailed level. The agreement between experiment and theory, in particular with such pronounced structures, demonstrates the full validity of the FC theory in the case of the dissociation of water in the first absorption band.

Population of Electronic Fine-Structure States

The FC theory with electronic fine structure predicts a selective population of both Λ-doublet and spin states. Because the selective population of Λ-doublet states is separately discussed in the next section, we concentrate here on some general aspects and on spin.

Although the FC theory predicts a weakly selective population of spin states, this is not supported by experiment. Within experimental uncertainty, all data seem to indicate a statistical population of spin states. This is true for room-temperature and jet-cooled species at 157 nm as well as for the room-temperature species at 193 nm and the true state-to-state data for the photolysis from single quantum states.

To demonstrate this behaviour, in Figure 9 the spin states in OH are analysed as to whether they are statistically populated for the state-to-state photodissociation from the 0_{00} level of H_2O. The population of spin states is statistical if, for a given rotational state N, the ratio of the populations in the multiplet states $^2\Pi_{\frac{1}{2}}$ and $^2\Pi_{\frac{3}{2}}$ is given by their statistical weights $N/(N + 1)$. Thus, a value of one in Figure 9 indicates a statistical population of spin states. Although there may be small deviations from one, a selective population of spin states cannot be concluded with the present signal-to-noise ratio. The corresponding theoretical points show almost no selective formation of spin states, in qualitative agreement with experiment. The spin selectivity is small in the dissociation of water in the first absorption band because OH is closer to Hund's case b than to Hund's case a for most states.

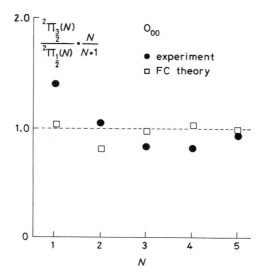

Figure 9 *Selective population of spin states. The ratio of the population in the multiplet state $^2\Pi_{\frac{3}{2}}$ to $^2\Pi_{\frac{1}{2}}$ is shown for different rotational states, multiplied by $N/(N + 1)$ to take into account the degeneracy. A value of 1 corresponds to statistical population*

A selective population of spin states has been found in several recent experiments.[47] In some cases the origin for spin selectivity has been ascribed to final-state interaction.[55] This may certainly be true; however, the evidence is not conclusive. Although we do not see any spin selectivity in the experiment under consideration, *the FC theory predicts in general non-statistical spin distributions*, in particular for molecules that are closer to Hund's case *a* and also for more linear parent molecules.[21]

This is an interesting effect, because the spin selectivity clearly originates from the first excitation step and not from the second scattering step. It can be washed out in the second step, depending upon the strength of final-state interaction. It seems less probable that the weak magnetic forces that couple spins can cause a considerable selectivity in the scattering step. Even in cases of strong rotational final-state interaction, the spin selectivity may originate from the first step and couple only subsequently to the rotational angular momenta, *i.e.* after the dissociation into the isolated products.

Selective Population of Λ-Doublet States

The most interesting aspect of the photodissociation of H_2O in the first absorption band is probably related to the selective population of Λ-doublet states. It is not only that the selective population of the upper Λ-doublet state explains the astronomical OH masers in regions of star formation but

[55] H. Joswig, M. A. O'Halloran, R. N. Zare, and M. S. Child, *Faraday Discuss. Chem. Soc.*, 1986, **82**, in press.

also that it can be very easily understood both qualitatively and quantitatively.

The highly selective population of Λ-doublet states is demonstrated by Figure 10, where the data of Figure 6 are plotted in a different way. The relative population of Λ-doublet states, *i.e.* for each rotational state N the ratio of the population in the upper and lower Λ-doublet states, is shown for the same three H_2O temperatures as before.

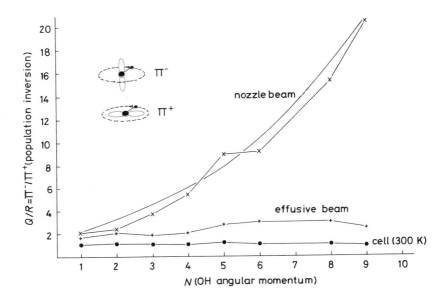

Figure 10 *Selective population of Λ-doublet states for the 157 nm photolysis of H_2O at different temperatures. The ratio of the population in the upper Π^- (⧂) to the lower Π^+ (∞) Λ-doublet state in $^2\Pi_{\frac{3}{2}}$ is shown as a function of the final OH rotational state. This ratio is obtained from the fluorescence intensities of Q and R lines in the LIF. The upper curve is for jet-cooled species in a nozzle beam, the middle for a slightly cooled effusive beam, and the lower for room-temperature photolysis*

For the jet-cooled H_2O a large preference for the upper Λ-doublet states (in $^2\Pi_{\frac{3}{2}}$) is found: for higher rotational states approximately 20 times more molecules are obtained in the upper than in the lower Λ-doublet state of OH. Although the preference depends strongly upon the final OH rotation, it is still a factor of two for the lowest rotational state $j = \frac{3}{2}$. It should be emphasized that a selective population of Λ-doublet states is surprising since these states are only about 0.05 cm^{-1} apart compared to an excess of energy of about 25 000 cm^{-1}!

The temperature dependence of the product-state distributions is seen again: the relative population of Λ-doublet states depends strongly upon the temperature of the parent molecule. An almost statistical behaviour is found for room-temperature H_2O, and a pronounced preference is observed for the slightly cooled species. This dominance is strong in the beam experiment. Again this is only possible if the state-to-state cross-sections depend upon the initial rotational state.

These experimental results lead to three important questions:

(a) What is the origin of the selective population of the upper Λ-doublet state?

(b) Why does the selectivity depend so strongly on the final OH rotation?

(c) Why does the selectivity depend so strongly on the H_2O temperature?

These questions will be answered in different ways. One way makes use of quantitative theory in which state-to-state cross-sections are calculated and averaged according to equation (5). This will be described in Section 7. It is, however, also possible to answer these questions in a qualitative, chemically intuitive way, and this will be done here.

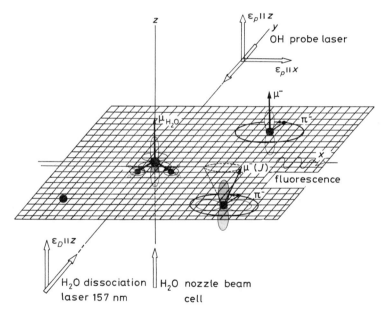

Figure 11 *Origin for the selective population of the upper Λ-doublet state of OH. The break-up of the excited complex into products OH and H is shown. The two different OH molecules that are shown are both for the upper Λ-doublet in $^2\Pi_\frac{3}{2}$. The upper part is the idealized high-j limit with the absorption dipole strictly perpendicular to the OH rotation plane. The lower part is for intermediate j, where both lobes are populated [see equation (25)]. In this case the orientation of the absorption dipole is j-dependent and precessing around the z-axis*

The selective population of Λ-doublet states is closely related to the conservation of electronic symmetry in the motion of heavy nuclei. This becomes obvious in Figure 11, which illustrates the break-up of H_2O to the products OH and H. The H_2O molecule defines the plane shown in the figure, and the unpaired $p\pi$ lobe in the excited complex is perpendicular to this plane. The fragmentation is completely planar because the $4a_1{}^*$ orbital, which is responsible for the strong forces during fragmentation, is symmetric to the plane. OH will therefore rotate in the identical plane. On the right-hand side two possible electronic configurations are shown for the OH product. The electronic configuration on the upper right-hand side is for the upper Π^- Λ-doublet state (see Figure 2); the lower electronic configuration is explained below. From this figure the qualitative origin for the selective population of the upper Λ-doublet state is obvious. The orientation of the unpaired $p\pi$ electron in the upper Λ-doublet is perpendicular to the OH rotation plane; for the lower Π^+ Λ-doublet it is in plane. In the excited complex there is also an unpaired electron perpendicular to the H_2O plane. Obviously the orientation of this unpaired electron is conserved in the course of the planar fragmentation.

The reason why the lobe orientation is conserved becomes immediately clear if we consider the reflection symmetry of the electronic wavefunction at the plane defined by the nuclei. Both the excited complex and the upper Λ-doublet are antisymmetric; the lower Λ-doublet is symmetric. Therefore, a conservation of the lobe orientation is equivalent to the *conservation of the electronic symmetry in the break-up of the excited complex*. Conservation of electronic symmetry is well known.

Electronic symmetry is used to classify different potential-energy surfaces; transitions among different surfaces are called non-adiabatic. The fragmentation of H_2O in the first absorption band proceeds along the asymmetric excited state $\tilde{A}^1B_1$. The selective population of the asymmetric Λ-doublet state is obtained because non-adiabatic transitions are small. The conservation of the lobe orientation simply reflects adiabatic behaviour. These qualitative arguments may be summarized in the simple correlation rule:

$$\text{8} \longrightarrow \text{8} \qquad \infty \longrightarrow \infty \qquad (24)$$

This is a very useful rule: an experimentally observed preference for the symmetric (asymmetric) Λ-doublet state implies that the process evolved via a symmetric (asymmetric) transition state.

This rule has been found to be qualitatively correct in many cases. Its importance becomes very clear if we compare the photodissociation of H_2O in the first absorption band with the chemical reaction of $O(^1D)$ with H_2.[56] In contrast to the photodissociation, which proceeds along the antisymmetric surface $\tilde{A}^1B_1$ and yields more OH in the antisymmetric Λ-doublet, the chemical reaction proceeds via the symmetric ground state $\tilde{X}^1A_1$ and yields more OH in the symmetric Λ-doublet state. In the break-up of the reactive complex an s–p bond breaks in the plane of the nuclei, leaving an unpaired $p\pi$ lobe in the rotation plane. Here, the close connection of the oriented $p\pi$ lobes to real chemistry becomes obvious: directed $p\pi$ lobes form

56 A. C. Luntz, *J. Chem. Phys.*, 1980, **73**, 1143.

strong chemical bonds and the selective population of Λ-doublet states reflects simply a memory of the chemical bond in the transition state. This type of explanation for the selective population of Λ-doublet states was discussed by Gwinn *et al.*[57] and has been used to explain the selective population of Λ-doublet states in many chemical reactions, inelastic collisions, and other photodissociation processes.

Although this simple chemical picture is extremely obvious, there was considerable doubt about it, which originated from a series of different attempts to explain astronomical OH masers.[58] All these attempts were based on the assumption that for the upper Λ-doublet the lobe was in plane. With this assumption the chemically intuitive picture was of course wrong. The photodissociation of H_2O in the first absorption band was the first instance in which a selective population of the *upper* Λ-doublet state in $^2\Pi_{\frac{3}{2}}$ (*i.e.* preference of Q lines in the LIF via the $^2\Sigma-^2\Pi$ absorption band!) was found. The selectivity was opposite to that found in $O(^1D) + H_2$, supporting the correlation rule (24). There is now no longer any doubt about the lobe orientation: it has been measured directly,[20] and the errors in theory have been admitted by most authors.[59,60]

Nevertheless the simple picture outlined above is only qualitatively correct. It predicts the population exclusively in one Λ-doublet state. However, all experiments in which selective population of Λ-doublet states is observed show only a *preference* for either the one or the other Λ-doublet. In contrast to the correlation rule (24), non-adiabatic transitions can occur if the Born–Oppenheimer approximation breaks down in the fragmentation process. In the parent H_2O the electrons are strongly coupled to each other whereas the coupling to the nuclear degrees of freedom is negligible. In the product OH this is clearly different. Both angular momenta of the unpaired electron couple to the nuclear rotation, according to Hund's coupling cases. Due to the coupling of electronic and nuclear motion the electronic symmetry, which is well defined in the parent, breaks down for the product; in the isolated OH there is no well defined electronic symmetry of the unpaired electron relative to the rotation plane. This implies that *a complete conservation of symmetry is impossible.*

This is also illustrated by the electron density distribution in the two Λ-doublet states: the nice picture with perfectly oriented lobes from Figure 2 is not realistic. Instead the electron density depends upon the rotational state, a clear manifestation of the breakdown of the Born–Oppenheimer approximation. This is illustrated by the two electronic configurations at the right of Figure 11. The upper right-hand area shows a perfectly oriented $p\pi$ lobe, which is obtained in the high-j limit. The lower right-hand area shows that both lobes are populated at lower and intermediate j. It can be shown

[57] W. D. Gwinn, B. E. Turner, W. M. Goss, and G. L. Blackman, *Astrophys. J.*, 1973, **179**, 789.
[58] D. P. Dewangan and D. R. Flower, *J. Phys. B*, 1983, **16**, 2157; R. N. Dixon and D. Field, *M.N.R.A.S.*, 1979, 583; H. Kaplan and M. Shapiro, *Astrophys. J. Lett.*, 1979, **229**, L91.
[59] J. L. Kinsey, *J. Chem. Phys.*, 1985, **81**, 6410.
[60] D. P. Dewangan and D. R. Flower, *J. Phys. B*, 1985, **18**, L137.

that the electronic-density distribution ρ for the unpaired electron is a superposition of the densities in the two $p\pi$ lobes perpendicular to the internuclear axis with j-dependent weights:[13]

$$\rho(\Pi^+) \sim (1 - c_j^2)\, \{ + c_j^2\, \infty$$

$$\rho(\Pi^-) \sim c_j^2\, \{ + (1 - c_j^2)\, \infty \tag{25}$$

with

$$c_j^2 = 0.5 + \left[4 + \frac{(\lambda - 2)^2}{(j - \frac{1}{2})(j + \frac{3}{2})} \right]^{-\frac{1}{2}} \tag{26}$$

$$\lambda = A/B$$

Here A is the multiplet splitting and B is the rotational constant. The parameter λ determines the j-dependence of the coefficients c_j^2, which can be very different for different types of molecules. Appreciable orientation effects are only obtained in Hund's case b, whereas for Hund's case a both lobes are equally populated.

In any case, c_j^2 will approach the value of one in the limit of large j.[13] In this 'high-j limit' the perfect orientation $\{$ is obtained for the Π^- and the perfect orientation ∞ is obtained for the Π^+ state (it should be mentioned that the $+/-$ notation on the Π symbol is opposite to that used earlier[13]). Only in the high-j limit is the electron density described by the perfectly oriented lobes in Figure 2, and *only in this high-j limit is a strict conservation of electronic symmetry possible*. A closer look at the very large preference for the $\{$ state at larger N in Figure 10 reveals that the symmetry is indeed almost completely conserved.

In a very simple model these j-dependent electron densities explain almost quantitatively the relative Λ-doublet population for the jet-cooled H_2O. The model is simply based upon the assumption that the population in the upper (lower) Λ-doublet is given by the probability for the out-of-plane lobe, $\{$. This is, according to equation (25), c_j^2 for the Π^- and $(1 - c_j^2)$ for the Π^+ Λ-doublet component. The relative population of Λ-doublet states is then simply given by $c_j^2/(1 - c_j^2)$, with no fit parameters in the model. Figure 10 demonstrates that this model works very well for the jet-cooled H_2O: the solid line is the result of this simple model. For warmer H_2O the model is not directly valid. However, it can be extended to include initial out-of-plane rotations of the parent molecule and then explain, at least qualitatively, the decreasing selectivity with temperature.[19]

For the detailed state-to-state experiments, however, the picture breaks down completely. Also the qualitative argument from correlation (24), which predicts a *consistent* preference for one Λ-doublet, is no longer valid. This was demonstrated in Figures 7 and 8.

These j-dependent electron densities lead to another interesting effect, which is important for the analysis of polarization experiments and has not been considered before. In Figure 11 the absorption dipoles (or, better, 'transition moments') for the electronic transitions in H_2O and OH are shown. For the $(\tilde{A}^1B_1 \leftarrow \tilde{X}^1A_1)$ transition in H_2O this dipole is perpendicular to the plane. For the transition $(^2\Sigma \leftarrow {}^2\Pi)$ in OH, which is used in the LIF, it is easy to show[19] that the dipole is always along the unpaired $p\pi$ lobe. This implies, for the high-j limit, that the dipole is either perpendicular to the OH rotation plane (as indicated by the arrow in the upper right-hand area in Figure 11) or parallel to the OH rotation plane. The in-plane dipole rotates with the OH because the lobe is strictly perpendicular to the moving internuclear OH axis. The dipole is, however, stationary if the lobe is perpendicular to the plane.

This leads to the behaviour shown at the lower right-hand side in Figure 11. As mentioned above, the electron density is distributed over both $p\pi$ lobes at low and intermediate j. Consequently, the absorption dipole has a stationary component perpendicular to the rotation plane and a rotating component in the rotation plane: *the resulting dipole is precessing with a j-dependent direction around the normal to the plane.* This has been ignored in the analysis of most polarization experiments. For $^2\Sigma-^2\Pi$ transitions (and probably for other cases as well) the precession will lead to a strong j-dependent polarization for both emission and excitation experiments. This j-dependent polarization can be easily misinterpreted as a dynamical effect.

A general consequence of the precession of these dipoles is that polarization is only expected for appreciable orientation of the unpaired $p\pi$ lobe, *i.e.* only for molecules with considerable Hund's case b character; in Hund's case a, *e.g.* NO at low j, no polarization is expected.

The Effect of the Bond Angle in the Parent Molecule

Figure 12 shows some model calculations for rotational and electronic fine-structure distributions, in which the bond angle γ_e of the parent molecule is *arbitrarily* varied. All parameters in these calculations are the same as before for H_2O. The only quantity that is varied is the bond angle γ_e of the parent molecule in the electronic ground state. Only dissociation of the rotational ground state 0_{00} is considered. Although these distributions will probably not be found in any realistic case, they demonstrate features that can be expected from the first step in photodissociation. Even if final-state interaction is strong for rotation, it may be small for electronic degrees of freedom and some memory of the first step may be found in the product-state distributions.

The distributions in Figure 12 demonstrate that the predictions of the FC limit with electronic fine structure are by no means trivial. The oscillations in the rotational distributions for a given Λ-doublet and multiplet state are most pronounced for $\gamma_e = 90°$. In this case the ground-state wavefunction is symmetric with respect to $90°$ and only even states ($j = 0, 2, \ldots$) contribute in expansion (15). The oscillations in the different Λ-doublet states are opposite to each other. Because both the Λ-doublet and the rotational states

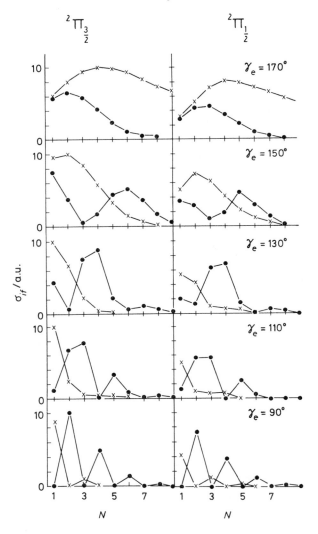

Figure 12 *Model calculations in the FC limit with electronic fine structure to study the effect of the bond angle of the parent molecule on the selective population of fine-structure states. All data are for the rotational ground state 0_{00}. The crosses are for the $\Pi^{+}-$ and the points for the $\Pi^{-}-\Lambda$-doublet state. The left-hand side is for the $^{2}\Pi_{\frac{3}{2}}$ and the right-hand side for the $^{2}\Pi_{\frac{1}{2}}$ multiplet state*

have different parity, this indicates that the oscillations originate from symmetry conservation in the dissociation process. For 110°, which is close to the actual case of H_2O, the oscillations are less pronounced, in particular at higher rotational states and for the $\Pi^{+}-\Lambda$-doublet. The $\Pi^{-}-\Lambda$-doublet is dominant at higher rotational states ($N \geqslant 2$), which is important for the selective population of Λ-doublet states. At 130° the structure remains almost unchanged but begins to shift to higher rotational states. At lower rotational states the Π^+–Λ-doublet state begins to dominate more strongly. This becomes even more pronounced at 150°. At 170°, which is close to the case of a linear molecule, the distribution is wide because many rotational states are required to describe the ground-state wavefunction. The lower Λ-doublet dominates for all rotational states. *The relative population of Λ-doublet and spin states depends in a very sensitive way on the angle of the parent molecule.*

These data demonstrate that in the FC limit, which describes the effect of the first step in photodissociation processes, results concerning the parent molecule in the ground state are yielded that are by no means trivial. The validity of the FC limit yields the additional information that the preferred angle in the excited state is similar to the angle in the ground state. A selective population of electronic fine-structure states can be expected from the first step in photodissociation, and memories of the first step may be found in the actual product-state distributions, even in cases with considerable final-state interaction.

6 Total-Absorption Cross-section and OH Vibrational Distributions

In Section 4 there was a discussion of how the completely resolved absorption cross-sections from a single state of H_2O can be approximated as a product of two terms, equation (23). One term describes the final OH rotational distribution including electronic fine structure. It can be fully described in the FC limit and is independent of the particular OH vibrational state and the photolysis wavelength. However, it depends strongly on the initial H_2O rotational state. The second term describes the final OH vibrational state. It is wavelength dependent and must be dynamically calculated including the full final-state interaction concerning the vibrational and translational coupling. To a large extent it is independent of the initial H_2O rotational state but depends strongly on the initial vibrational level.[34] The rotational part of the absorption cross-section has been discussed in Section 5, from both an experimental and a theoretical point of view. The vibrational part is the topic of this section.

Calculation

The cross-section for absorbing a photon with wavelength λ and producing OH in a particular vibrational state n is calculated under the following assumptions and approximations. First, the electronic fine structure of OH is completely ignored. Second, only dissociation of the rotational–vibrational ground state of H_2O is considered. Because of the two (equivalent) dissociation channels the determination of Ψ_{ex} is by far the most difficult part of the dynamical calculation, and an exact solution of the equations of motion is not possible, even if electronic degrees of freedom are neglected.

As mentioned in Section 4, the so-called energy sudden approximation[61] (ESA) is used in the calculation of Ψ_{ex}. Within the ESA it is assumed that the rotational energy of OH is negligibly small compared to the total available energy *throughout* the dissociation. The ESA is perfectly justified in the dissociation of water in the first absorption band. Around the ground-state equilibrium, $\gamma_e \approx 104°$, where trajectories are started, the torque $\partial V_{ex}/\partial\gamma$ exerted on OH is very weak at all H—OH internuclear distances, with the consequence that *j* remains small throughout the trajectory. The ESA has been rigorously tested by comparison with a more exact treatment; however, one of the OH distances was fixed in these calculations.[25] Usually the rotational angular momentum *j* is assumed to be zero within the ESA and then the orientation angle γ is a constant of motion as it readily follows from Hamilton's equations (8). Thus, within the ESA the dimensionality of the classical and quantum mechanical equations of motion is significantly reduced. However, the fixed orientation angle enters all equations as a parameter through the interaction potential.

The calculations are then performed in the following way. First, partial cross-sections $\sigma_n(\lambda|\gamma)$ are determined for fixed orientation angle γ. The fully three-dimensional cross-sections are finally obtained by averaging over γ, according to:

$$\sigma_n(\lambda) \sim \int_0^\pi d\gamma \sin\gamma \sigma_n(\lambda|\gamma) \tag{27}$$

The partial cross-sections are calculated quantum mechanically from the Golden Rule expression:[28]

$$\sigma_n(\lambda|\gamma) \sim v \, | < \Psi_{gr}(R,r,\gamma) | \mu_\varepsilon(R,r,\gamma) \, | \Psi_{ex}^n(R,r|\gamma) > |^2 \tag{28}$$

Here, Ψ_{gr} is the nuclear wavefunction in the ground electronic state and μ_ε is the component of the transition dipole function in the direction of the polarization of the electric vector. Ψ_{ex}^n is the approximate dissociative wavefunction with an outgoing plane wave in vibrational channel *n*. It depends parametrically on the orientation angle γ. The integration in equation (28) is over the stretch co-ordinates *R* and *r*. Since the ground-state wavefunction is restricted to a narrow angular interval, only angles $|\gamma - \gamma_e| < 20°$ with $\gamma_e \sim 104°$ have to be considered in equation (27).

The evaluation of $\Psi_{ex}^n (R,r|\gamma)$ is not performed by using the normal Jacobi co-ordinates *R* and *r* but by employing two-dimensional polar co-ordinates (hyperspherical or Delves co-ordinates).[62] These co-ordinates are now routinely used in collinear reaction studies because they have certain advantages concerning the numerical treatment. The final results are, how-ever, independent of the particular sets of co-ordinates used in the calcu-lation. The two-dimensional scattering problem is treated exactly by expand-ing Ψ_{ex}^n into suitable basis functions. Details of the numerical treatment are fully described in reference 34.

[61] D. Secrest, *J. Chem. Phys.*, 1975, **62**, 710.
[62] J. Manz, *Comments At. Mol. Phys.*, 1985, **17**, 91.

The energy level and the bound wavefunction for the ground electronic state, $\tilde{X}^1 A_1$, are calculated *exactly* using the empirical-fit potential of Sorbie and Murrell.[54] The energy of the (0,0,0) 0_{00} ground state is -9.500 eV, relative to the H + H + O limit. The excited-state potential, $V_{ex}(\tilde{A}^1 B_1)$, is fixed to the ground state at infinite H—OH separation. This fixing is necessary because $V_{gr}(\tilde{X}^1 A_1)$ and $V_{ex}(\tilde{A}^1 B_1)$ are taken from completely different sources. Besides that, the excited-state potential is used *without any modification*.

The $\tilde{A} \leftarrow \tilde{X}$ transition dipole function is calculated in the Franck–Condon region using SCF electronic wavefunctions.[34] It has a rather weak coordinate dependence and affects only slightly the variation of the cross-section with wavelength.[34]

In conclusion, the dynamical calculation contains an empirical potential for the ground state, a calculated excited-state potential, and a calculated transition dipole function. The only approximation is the ESA for the rotational degree of freedom of the nuclear wavefunction in the excited state, which, however, can be considered to be exact for the dissociation of water in the first absorption band. A similarly complete treatment does not exist at present for any other system.

The Total-Absorption Cross-section

Figure 13 shows the calculated total-absorption cross-section $\sigma_{tot}(\lambda) = \Sigma_n \sigma_n(\lambda)$ *versus* photon wavelength λ for dissociation of the (0,0,0) 0_{00} ground state. It is rather broad without any sharp structures, as expected for a fast and direct photodissociation process. The theoretical curve is compared to the experimental cross-section for dissociation of water initially at room temperature.[16] Theory and experiment are normalized to each other at the maximum around $\lambda \approx 1650$ Å. In order to facilitate the comparison, the experimental base line[16] is added to the theoretical cross-section.

This is the first direct comparison between experiment and *ab initio* theory for the photodissociation of a triatomic molecule. The agreement is astonishingly good. The calculation does not include any adjustable parameter. Even the vague undulations superimposed on the experimental cross-section are well reproduced.

The small deviation at the red side of the spectrum is very likely due to the omission of higher initial total angular momentum states in the calculation. Since the time taken to compute $\sigma_{tot}(\lambda)$ for all rotational states populated at room temperature is prohibitively long, the Boltzmann average was incorporated in the following simple way. First, it is assumed that the total cross-section as a function of the scattering energy, $E = E_i + h\nu$, is independent of the particular rotational level of H_2O. Then, on the wavelength scale each cross-section is shifted to the red according to the energy shift $|E(J_{K_aK_c}) - E(0_{00})|$. Finally, the Boltzmann average for $T = 300$ K is performed on the λ-axis.

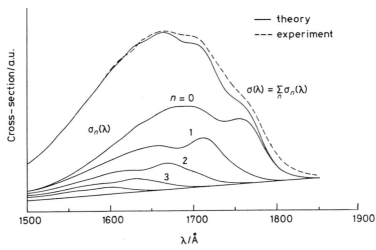

Figure 13 *Comparison of the theoretical and the experimental[16] total cross-section $\sigma(\lambda)$ for dissociation of the (0,0,0) vibrational state (linear scale). The experimental base line[16] is added to the theoretical cross-section. Theory and experiment are normalized at the maximum. The theoretical cross-section is calculated for $J = 0$; the experimental cross-section is measured at 300 K. The partial cross-sections $\sigma_n(\lambda)$ are also shown for dissociation into specific vibrational OH states, n*

The results for the dissociation of H_2O and D_2O are compared with experiment in Figure 14. The agreement is almost perfect. The spectrum for D_2O is slightly shifted to the blue side simply because the ground-state energy for D_2O is below that for H_2O. It is also narrower than the H_2O spectrum. This follows from the fact that the radial wavefunction for the same scattering energy oscillates faster for D_2O because of the larger reduced mass.

The experimental total-absorption cross-sections in Figures 13 and 14 show some poorly resolved structures that are also reproduced by the calculation.[24,34] In reference 16 these undulations are explained as selective absorption (pre-dissociation) due to the bending motion on the excited-state potential surface. The calculation clearly demonstrates that this interpretation is wrong. The bending motion in the dissociative state is treated in the sudden limit, *i.e.* the bending angle is fixed, and therefore cannot lead to selective absorption. The partial cross-sections $\sigma_n(\lambda|\gamma)$ in equation (28) show the same structures and there they are even more pronounced.

Figure 13 clearly shows that the undulations in the total cross-section originate from the structures in the vibrationally resolved cross-sections $\sigma_n(\lambda)$. The latter can be explained on the basis of the $\tilde{A}^1B_1$ potential surface, especially its co-ordinate dependence within the barrier region.[24,34] The summation of several partial cross-sections $\sigma_n(\lambda)$, which are gradually shifted to the blue, leads to the sequence of structures in the total-absorption cross-section.

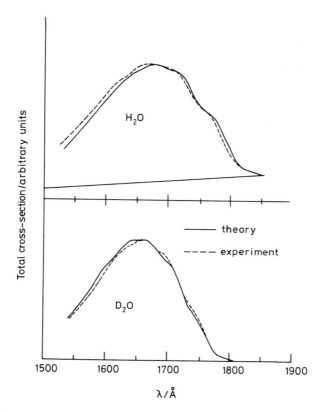

Figure 14 *Comparison of the theoretical, Boltzmann-averaged, and experimental*[16] *total cross-sections* $\sigma(\lambda)$ *for dissociation of the* (0,0,0) *vibrational state of* H_2O *and* D_2O *(linear scale). The experimental base line*[16] *for* H_2O *is added to the theoretical cross-section. It is essentially zero for* D_2O. *Theory and experiment are normalized at the maxima*

The maximum of the absorption cross-section and its width depend very sensitively on the excited-state potential in the transition region, primarily on the vertical excitation energy and the variation of V_{ex} with the dissociation co-ordinate. This underlines the high accuracy of the calculated $\tilde{A}^1B_1$ potential-energy surface. On the dynamical side it must be emphasized that a 'reactive', *i.e.* two-channel, treatment is essential. An 'inelastic' calculation considering only one dissociation channel, as previously done to describe process (1),[23] is wrong.

Vibrational Distributions

Also shown in Figure 13 are the partial-absorption cross-sections $\sigma_n(\lambda)$ for dissociation into the nth vibrational state of OH. The total-absorption cross-section is the sum of all partial cross-sections. The individual vibrational cross-sections all have roughly the same shape. However, they are successively shifted to the blue with increasing vibrational state. This shift approximately correlates with the energy levels of the symmetric stretch vibrational mode in the excited state. It does not correspond to the difference in the vibrational energies of the free OH molecule. The wavelength dependence of these partial cross-sections has not yet been measured.

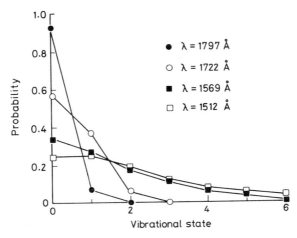

Figure 15 *Theoretical vibrational-state distributions for dissociation of the (0,0,0) vibrational state of water for several wavelengths*

The data for the partial cross-sections in Figure 13 are plotted in Figure 15 as vibrational-state distributions for different wavelengths. These distributions become gradually broader as λ decreases. They are exclusively determined by final-state interaction, *i.e.* they cannot be explained in the FC limit. Since the OH equilibrium separation within the H_2O molecule is roughly equal to the equilibrium in the free molecule, the FC theory would predict that almost all products are channelled into the $n = 0$ ground state, *independent of photon wavelength*. By analogy to equation (8), the time derivative of the classical (continuous) vibrational quantum number, $n(t)$, is proportional to:

$$\frac{dn}{dt} \sim \frac{\partial V_1(r)}{\partial r} \tag{29}$$

where $V_1(R,r) = V(R,r) - v(r)$ is the interaction potential and $v(r)$ is the potential function for the free molecule. It is easy to surmise from Figure 4 that this force, acting on the oscillator, is strong and increases with increasing energy. It is therefore not surprising that the final vibrational distribution becomes successively broader with increasing energy. The usual argument, which predicts a 'cold' vibrational distribution if the equilibrium separation does not change significantly during dissociation, is simply wrong in this case. It ignores the influence of the excited-state potential, especially its dependence on the vibrational co-ordinate.[63]

At 157 nm the ratio of OH molecules produced in the vibrational states $n = 0, 1$, or 2 is $1.0 : 1.0 : 0.58$. It should be mentioned that the value for $n = 2$ from reference 19 has been corrected for predissociation according to reference 64. Higher levels than $n = 2$ could not be detected by LIF because of predissociation in the electronically excited state of OH. The corresponding theoretical ratio is $1.0 : 0.81 : 0.53$, in good agreement with experiment. However, a real test of the theoretical prediction would only be possible if the complete vibrational distribution at several wavelengths could be measured. Such measurements are currently in progress.[65]

7 Averaging and Synthesis of Temperature-Dependent Product-State Distributions

Only in very rare cases have state-to-state photodissociation cross-sections been measured. In almost all experiments some averaging over initial rotational states is involved. The typically measured quantities are the temperature-dependent product-state distributions [equation (5)] or the total-absorption cross-section [equation (6)], which result from averaging state-to-state cross-sections with the corresponding Boltzmann weights.

The knowledge of state-to-state photodissociation cross-sections allows the synthesis of the product-state distributions *at all different temperatures*, which would certainly be impossible to measure otherwise. Even more complex non-equilibrium situations, which are not unusual in astrophysics, can be modelled. However, it should be emphasized that *all* state-to-state cross-sections $\sigma_{if}(\lambda)$ have to be known, *i.e.* for all wavelengths λ, all initial states i, and all final states f. For the case of H_2O these cross-sections have been determined for all rotational states of the parent H_2O with $J_i \leqslant 6$ and all $2J_i + 1$ substates, for all OH states (excluding nuclear spin and magnetic sublevels), and all wavelengths.

Averaging over Λ-Doublets

As a first example, averaging over only the two Λ-doublets is considered. Figure 16 shows rotational distributions for the dissociation of the 0_{00}, 4_{04}, 4_{14}, and 4_{31} states of H_2O, which are averaged (summed) over the lower and upper Λ-doublets. The data are given in a Boltzmann plot. The surprising result is that, in contrast to the pronounced structures in the original distributions (Figures 7 and 8), all the points lie on a straight line: the Λ-doublet averaged distributions can be described by a Boltzmann temperature! The high selectivity is already washed out by averaging over only one degree of freedom.

[63] S. Hennig, V. Engel, and R. Schinke, *J. Chem. Phys.*, 1986, **84**, 5444.
[64] K. German, *J. Chem. Phys.*, 1975, **63**, 5252.
[65] K. H. Welge, private communication.

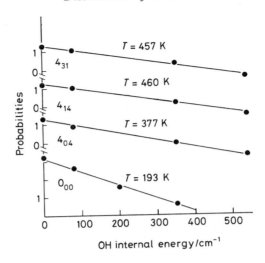

Figure 16 *Experimental OH rotational-state distributions for dissociation of single rotational states of H_2O as indicated. The results are averaged over the two Λ-doublet states in $^2\Pi_{\frac{3}{2}}$. The lines indicate Boltzmann distributions with temperatures given for each initial state*

The rotational temperatures that are obtained from the Boltzmann plot still depend upon the initial state of the parent molecule. For the rotational ground state the temperature is much lower than for the rotationally excited states with $J_i = 4$. Obviously more and higher OH rotational states are required to describe the higher rotational states of the parent molecule in the expansion (15) or (20). The projection quantum numbers K_a and K_c cause only a slight effect, in contrast to Figures 7 and 8. However, there is some difference in the rotational temperatures for the states 4_{04} and 4_{31}.

Synthesis of the Selective Population of Λ-Doublet States

Here the quantitative answer is given to the questions asked in Section 5 about the selective population of Λ-doublet states. Figure 17 shows how the selective population of Λ-doublet states, seen in Figure 10, arises from averaging over different initial quantum states. The theoretical state-to-state cross-sections are used in the Boltzmann average according to equation (5) for several H_2O temperatures.

The result for 10 K shows a highly selective Λ-doublet population in favour of the upper Λ-doublet in $^2\Pi_{\frac{3}{2}}$. The selectivity increases with increasing OH rotation. Superimposed on the general increase is a strong structure which demonstrates that there is still a memory of the pronounced structure of the state-to-state cross-sections. For H_2O temperatures above 40 K the relative population of Λ-doublets is already much smoother, more like the experimental data of Figure 10. For 100 K the selectivity starts to decrease,

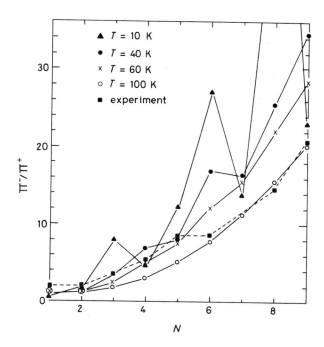

Figure 17 *Synthesis of the selective population of Λ-doublet states. The relative population of Λ-doublet states is shown for different N. The symbols are described in the inset. The experimental data are for the nozzle beam*

and it is almost completely lost at 300 K. The corresponding experimental distribution for the jet-cooled species is also given in Figure 17. A comparison with the averaged data shows the very good agreement between experiment and theory. All the questions that were qualitatively discussed before are answered quantitatively here: the preference for the upper Λ-doublet state in $^2\Pi_{\frac{3}{2}}$, the increasing selectivity with increasing final OH rotation, as well as the decreasing selectivity with increasing H_2O temperature. The theoretical data suggest a temperature of ~ 100 K for the H_2O nozzle beam.

Synthesis of Rotational Distributions

As a last example of averaging, the question of how the rotational-energy content in the OH product is composed from the different state-to-state cross-sections is considered. Table 1 shows the rotational-energy content in the product OH at different H_2O temperatures. The H_2O temperature, $T(H_2O)$, and the 'rotational temperature' of OH, $T(OH)$, are given in both the upper (Π^-) and the lower (Π^+) Λ-doublet states of $^2\Pi_{\frac{3}{2}}$. The term rotational temperature is given in parentheses because the distributions are considerably non-Boltzmann. They are obtained by a least-squares fit together with the standard deviations ΔT, which are also given in Table 1.

Table 1 *Product* OH *'rotational temperature' as a function of parent* H_2O *temperature*

$T(H_2O)/K$	\[$T(OH) \pm \Delta T$\]/K	
	Π^-	Π^+
10	413 ± 101	168 ± 25
50	393 ± 10	185 ± 13
150	482 ± 2	280 ± 3
200	522 ± 3	328 ± 4
250	556 ± 4	370 ± 5
300	585 ± 5	408 ± 6

ΔT is a measure of the deviation from Boltzmann behaviour. It should be mentioned that the summation over state-to-state cross-sections includes only H_2O rotational states up to $J_i = 6$. The Boltzmann average, therefore, is converged only for temperatures below 150 K. The contribution from higher rotational states, which are not considered here, will tend to increase the OH rotational temperature, especially for higher H_2O temperatures. Therefore the rotational temperatures in the products are too low as compared to experiment.

Table 1 clearly demonstrates that the rotational energy in the product increases with increasing rotational energy in the parent. Both the rotational temperature and the increase in the rotational temperature are different for the two Λ-doublet components. It should also be mentioned that the structures in the original state-to-state cross-sections are immediately washed out even if only a few initial states are averaged. This is demonstrated by the rapid decrease of ΔT with increasing $T(H_2O)$.

8   Astronomical Masers

The orientation of the unpaired $p\pi$ electron in OH turns out to be very important for the explanation of astronomical OH masers, simply because these masers operate between the Λ-doublet states. Processes that selectively populate the upper Λ-doublet state in OH generate inversion and can explain these masers. The photodissociation of H_2O in the first absorption band is the one and, up to now, only process that yields a selective population of the upper Λ-doublet state in $^2\Pi_{\frac{3}{2}}$, and therefore the only process that explains the strong interstellar OH masers. Before the present experiment it was generally accepted that the $p\pi$ lobe was in plane for the upper maser level. This is why processes that proceed via symmetric transition states in the sense of the correlation rule (24) had previously been proposed as explanations for OH masers. Two such examples were the reaction of H with H_2O and the inelastic collision of OH with H_2. With the reversed orientation of the $p\pi$ lobe these processes lead to a selective population of the *lower* Λ-doublet state. This has been proved directly in recent experiments,[57,66] thus eliminating these suggested pump mechanisms.

[66] P. Andresen, D. Häusler, and H. W. Lülf, *J. Chem. Phys.*, 1984, **81**, 571.

Because the lobe orientation is so important, it was measured in a polarization experiment.[19]

The pumping of interstellar OH masers by the photodissociation of H_2O in the first absorption band is extremely obvious. It is surprising that it has not been considered before, because (*a*) the interstellar OH masers are found in regions of new-born stars that emit up to 30% of their luminosity in the range of the first absorption band and (*b*) it is generally assumed that the high OH abundance in these regions originates from photodissociation of H_2O in the first absorption band. The OH is formed inverted between the maser levels in the first step of its life. However, the quantitative modelling of astronomical masers has still to be done. It requires the complete knowledge of the state-to-state cross-sections, which are derived in this chapter.

9 Conclusions

The photodissociation of H_2O in the first absorption band is one of the best studied processes and turns out to be a model system for direct photodissociation. The elaborate studies described in this chapter were possible because a particularly clear physical situation is met, essentially due to the fact that the dissociation proceeds along one single excited-state potential surface that is well separated from all other states. This allows not only a qualitative understanding of the basic principles of direct dissociation processes but also a quantitative understanding in terms of theory. Both the experimental and theoretical studies represent the state of the art for current studies of fragmentation processes.

The OH product-state distributions have been analysed experimentally in great detail for a few wavelengths (157 nm, 193 nm, and i.r. + 193 nm) and several initial conditions of the parent molecule (room temperature, jet cooled, single quantum states). The vibrational and rotational distributions indicate a decoupling of vibrational and rotational motion.

The product-state distributions for both rotational and electronic fine-structure degrees of freedom are perfectly described by the Franck–Condon limit. This follows directly from the small anisotropy of the excited-state potential-energy surface in the region of the ground-state equilibrium geometry. The $\tilde{A}^1B_1$ state prefers the same bond angle as the $\tilde{X}^1A_1$ ground state. The product-state distributions are governed by quantum mechanics and depend sensitively on the initial state of the parent molecule. They cannot be described by classical mechanics, as in many other cases. The measured cross-sections represent directly the expansion coefficients of the ground-state wavefunction in terms of product-state eigenfunctions. The inclusion of the electronic fine structure in the FC theory, which has not been done previously for triatomic molecules, leads to quantitative agreement between experiment and theory. Although the product-state distributions are governed by the FC limit, they are by no means trivial. This is true in particular for the selective population of Λ-doublet states.

The general trend for the rotational degree of freedom is that all the distributions are 'cold', *i.e.* only low rotational states are populated. This is easily understood by the negligibly small torque acting on OH in the upper state. Only those rotational states are populated that contribute to the bending motion in the parent H_2O.

Although theory predicts a small effect for the selective population of spin states, this is not found experimentally. Even on a state-to-state basis the spin states seem to be populated statistically. However, the experimental uncertainties are too big to exclude a slight preference for one or the other spin state.

The selective population of Λ-doublet states is certainly one of the most interesting aspects of the results described in this chapter. The selectivity is found for internally 'cold' water. It favours the upper Λ-doublet and thus explains the astronomical OH masers in regions of star formation. For the temperature-dependent product-state distributions the preference was always for the same antisymmetric Λ-doublet state. For the state-to-state data, the preference depends upon the final state of OH *and* the initial rotational state of H_2O. Only considerable averaging over various rotational states of the parent molecule gives the consistent preference for the population of the upper Λ-doublet.

The preference for the more asymmetric Λ-doublet state can be qualitatively explained in a convenient, intuitive way by the conservation of symmetry during the break-up of the excited complex. However, on the state-to-state level the simple intuitive picture breaks down. The fact that the state-to-state cross-sections are given by expansion coefficients of the ground-state wavefunctions in terms of complicated product-state eigenfunctions leads to the strongly structured product-state distributions.

The photodissociation of H_2O in the first absorption band is a limiting case of a direct dissociation with a very small final-state interaction for rotation and electronic fine-structure degrees of freedom. This gives the only example up to now in which the features of the first step in photodissociation can be seen directly. It is a quantitative example for the FC limit, and the complicated data demonstrate that the features found in the FC limit are by no means trivial.

The final vibrational distributions cannot be described in the FC limit but are governed by strong vibrational–translational coupling in the exit channel. The calculated vibrational distributions become gradually broader with decreasing wavelength. This is in accord with the limited experimental data presently available.

The calculated total-absorption cross-sections for H_2O and D_2O agree almost exactly with experiment. These calculations do not contain any fit parameter, which underlines the high accuracy of both the *ab initio* energy surface for the $\tilde{A}^1B_1$ state and the dynamical calculation. We are now able to predict reliably quantities that are difficult to measure, for example the wavelength dependence of the absorption cross-section for a particular OH vibrational level or the absorption cross-section for dissociation of excited vibrational states of H_2O.

Finally, a complete description of the photodissociation of H_2O in the first absorption band is given on a state-to-state basis, which holds for all degrees of freedom.

Acknowledgements. We are very grateful to our colleagues G. G. Balint-Kurti, V. Engel, D. Häusler, E. W. Rothe, G. S. Ondrey, and V. Staemmler, who contributed significantly to this study over the past five years.

CHAPTER 4

High-Resolution Photochemistry: Quantum-State Selection and Vector Correlations in Molecular Photodissociation

M. P. DOCKER, A. HODGSON, and J. P. SIMONS

1 Introduction

An earlier review[1] listed some guidelines for the experimentalist planning ever more penetrating studies of the dynamics of molecular photodissociation. High on the list of recommendations were 'initial parent molecular-state selection' and 'measurements of the anisotropy of vector properties such as the photofragment angular distributions, rotational alignment, and orientation'.

The ability to achieve full initial quantum-state selection is important, since it allows the experimentalist to take (at least partial) active control over the initial conditions of the scattering process, rather than just be a passive observer of an ill-defined event. When the ability to choose alternative initial quantum states in the parent molecule is also combined with spectroscopic analysis of the distribution over the final quantum states in the scattered fragments, molecular photodissociation provides a uniquely detailed model system for the study of the state-to-state influence of angular momentum in the parent molecule on the photodissociation dynamics.

Angular momentum distributions and anisotropies in the scattered fragments have been probed through studies of the photofragment fluorescence polarization, through the probe polarization dependence of laser-induced fluorescence (LIF), through measurements of angle-resolved photofragment

---

[1] J. P. Simons, *J. Phys. Chem.*, 1984, **88**, 1287.

time-of-flight distributions following pulsed-laser photodissociation, and, most recently, through analysis of the Doppler-broadened profiles of individual features in their absorption spectra. Historically, the first to be measured were the angle-resolved velocity distributions that reflect the vector correlations (v, μ), where v and μ are, respectively, the relative velocities of the recoiling fragments and the parent molecular transition moment.[2] Subsequent measurements of the polarized spontaneous fluorescence and probe polarization dependence of LIF from diatomic photofragments provided a second set of vector correlations (j, μ), involving the photofragment rotational angular momentum j.[3-6] During the past year the triangle of vector correlations has been completed through analysis of the Doppler profiles of lines in the LIF excitation spectra of recoiling molecular fragments, which are sensitive not only to the correlation between j and μ and between v and μ but also to the rotational–translational correlation between v and j and between v, j, and μ.[7-14] In a triatomic molecular photodissociation in the high-j limit, for example, the rotating diatomic fragment necessarily recoils with its rotation vector j lying perpendicular to the relative velocity vector v.[8,9] Where there are more than three atoms and the molecule is non-planar, the existence of any correlation (v, j) becomes non-trivial. More importantly this correlation, unlike the first two, is referenced not to the laboratory frame but to the body frame, since μ and hence E are not involved; the correlation can persist therefore, even when there is a long interval before dissociation. In this sense it may be regarded as the correlation that most directly reflects the exit channel dissociation dynamics.[7]

The two or three years that have elapsed since the earlier review was written[1] have seen the goal of full quantum-state selection achieved and the triangle of vector correlations completed. The pace of experimental innovation and the accompanying theoretical analysis has been so intense as to leave at least one of the present authors (the oldest one) breathless: it clearly calls for an update focused on just these two topics. Section 2 reviews recent studies of predissociation from rovibronically selected levels, particularly water in the $\tilde{C}^1B_1$ Rydberg state via the second absorption continuum,

[2] K. R. Wilson in 'Excited State Chemistry', ed. J. N. Pitts, Gordon and Breach, New York, 1970.
[3] M. T. Macpherson, J. P. Simons, and R. N. Zare, *Mol. Phys.*, 1979, **38**, 2049.
[4] C. H. Greene and R. N. Zare, *Annu. Rev. Phys. Chem.*, 1982, **33**, 119.
[5] C. H. Greene and R. N. Zare, *J. Chem. Phys.*, 1983, **78**, 6741.
[6] R. Altkorn and R. N. Zare, *Annu. Rev. Phys. Chem.*, 1984, **35**, 265.
[7] R. N. Dixon, *J. Chem. Phys.*, 1986, **85**, 1866.
[8] G. E. Hall, N. Sivakumar, P. L. Houston, and I. Burak, *Phys. Rev. Lett.*, 1986, **56**, 1671.
[9] G. E. Hall, N. Sivakumar, R. Ogorzalek, G. Chawla, H.-P. Haerri, P. L. Houston, I. Burak, and J. W. Hepburn, *Faraday Discuss. Chem. Soc.*, 1986, **82**, in press.
[10] M. P. Docker, A. Hodgson, and J. P. Simons, *Chem. Phys. Lett.*, 1986, **128**, 264.
[11] M. P. Docker, A. Hodgson, and J. P. Simons, *Faraday. Discuss. Chem. Soc.*, 1986, **82**, in press.
[12] S. Klee, K.-H. Gericke, and F. J. Comes, *J. Chem. Phys.*, 1986, **85**, 40.
[13] K.-H. Gericke, S. Klee, F. J. Comes, and R. N. Dixon, *J. Chem. Phys.*, 1986, **85**, 4463.
[14] M. Dubs, U. Brühlmann, and J. R. Huber, *J. Chem. Phys.*, 1986, **84**, 3106.

$\tilde{B}^1 A_1$, and of ammonia in the $\tilde{A}^1 A_2''$ state; the discussion complements that in Chapter 3. In Section 3 we review some of the new experiments conducted by the groups in Zürich (Huber),[14] Cornell (Houston),[8,9] Frankfurt (Comes),[12,13] and Nottingham[10,11] and the accompanying theoretical framework formulated by Dixon,[7] which together demonstrate and develop the application of Doppler-broadened spectral analysis for the determination of photofragment vector correlations and dissociation dynamics. In Section 4 we attempt to identify a few more promising directions for the rest of the decade.

2 Quantum-State-Selected Photodissociation

Electronic transitions in polyatomic molecules, particularly those involving non-bonding electrons promoted into Rydberg orbitals, commonly excite quasi-bound states.[15] Where these are only weakly coupled into neighbouring dissociation continua, the excited-state lifetime may be sufficient to allow the resolution of individual rovibronic levels, particularly when the molecule is a hydride and the level spacing is not unduly congested. Under such conditions, the absorption of tunable, monochromatic, and polarized photons can allow full definition of the initially populated quantum states. Their subsequent evolution can be probed by resonance-enhanced multi-photon ionization (REMPI), by spontaneous or stimulated photon emission (the ion 'dip' or fluorescence 'dip' experiment[16,17]), and/or by spectral analysis of the primary photofragments (see Figure 1). Only in the last case is the dissociation followed to its conclusion; in the other techniques the focus is on the parent molecule itself and the initial stage of the dissociation – competing against the process that is experimentally monitored. Two molecules that have been very thoroughly studied in this way are water and ammonia and their fully deuteriated analogues. Our discussion will centre on these. We apologise for neglecting other examples, especially the beautifully detailed studies of state-selected photodissociation from intravalence states of $NCNO$[18,19] and H_2CO[20-22] by Wittig and by Bradley Moore and their co-workers; the former are reviewed thoroughly in Chapter 5.

[15] M. N. R. Ashfold, M. T. Macpherson, and J. P. Simons, *Top. Curr. Chem.*, 1979, **86**, 1.

[16] J. Xie, G. Sha, X. Zhang, and C. Zhang, *Chem. Phys. Lett.*, 1986, **124**, 99.

[17] M. N. R. Ashfold, C. L. Bennett, and R. N. Dixon, *Faraday Discuss. Chem. Soc.*, 1986, **82**, in press.

[18] C. X. W. Qian, M. Noble, I. Nadler, H. Reisler, and C. Wittig, *J. Chem. Phys.*, 1985, **83**, 5573.

[19] S. Buelow, M. Noble, G. Radhakrishnan, H. Reisler, C. Wittig, and G. Hancock, *J. Phys. Chem.*, 1986, **90**, 1015.

[20] D. J. Bamford, S. V. Filseth, M. F. Foltz, J. W. Hepburn, and C. B. Moore, *J. Chem. Phys.*, 1985, **82**, 3032.

[21] D. Debarre, M. Lefebvre, M. Péalat, J.-P. E. Taran, D. J. Bamford, and C. B. Moore, *J. Chem. Phys.*, 1985, **83**, 4476.

[22] H. Bitto, D. R. Guyer, W. F. Polik, and C. B. Moore, *Faraday Discuss. Chem. Soc.*, 1986, **82**, in press.

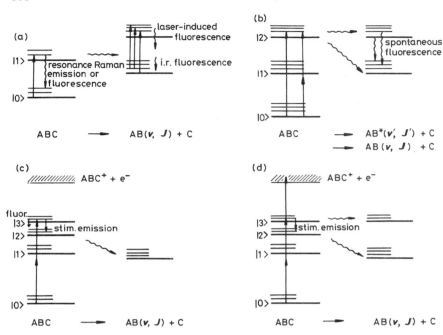

Figure 1 *Some typical photodissociation schemes.* (a) *Pump (photodissociation) and probe (laser-induced photofragment fluorescence); alternatively spontaneous resonance fluorescence or resonance Raman emissions from parent molecule. Also note CARS photofragment probe technique.* (b) *Pump (single- or two-photon photodissociation) and probe (spontaneous or laser-induced fluorescence).* (c) *Pump (two-photon excitation to the upper state $|3>$) and probe (via stimulated emission into selected predissociating levels in $|2>$); monitor predissociation spectrum/rates via changes in spontaneous $|3> \rightarrow |2>$ fluorescence intensity – fluorescence 'dip' technique.* (d) *Pump (two- or three-photon excitation into predissociating levels in $|3>$) and probe (via sequential REMPI). Alternatively probe predissociation in intermediate state $|2>$, via changes in ion current as $|3> \rightarrow |2>$ transition is stimulated – ion 'dip' technique*

H_2O *and* D_2O

The first rotationally structured band system in H_2O and D_2O lies near 124 nm and is associated with the origin of the $\tilde{C}^1B_1 \leftarrow \tilde{X}^1A_1$ electronic transition.[23] It is populated through promotion of a non-bonding electron from the out-of-plane orbital $1b_1$ into the Rydberg orbital $3pa_1$. Although the excited state retains very nearly the same equilibrium geometry as the ground state, its rovibronic levels are broadened by predissociation, more strongly in H_2O than in D_2O.[23] The broadening is rotationally state dependent, indicating the operation of a heterogeneous predissociation

[23] J. W. C. Johns, *Can. J. Phys.*, 1963, **41**, 209; 1971, **49**, 944.

pathway; the bending potentials shown in Figure 2 suggest predissociation via the $\tilde{B}^1A_1$ state, the state associated with the second absorption continuum generated by promotion of the inner lone-pair electron $3sa_1 \leftarrow 3a_1$. This promotion allows the bond angle to open towards the linear configuration where the $\tilde{B}^1A_1$ and $\tilde{A}^1B_1$ states become degenerate: they are Renner–Teller components of linear $H_2O(^1\Pi_u)$. The rapid change in bond angle following radiative or radiationless transfer onto the $\tilde{B}^1A_1$ surface generates a powerful torque as the H—OH distance increases and the excited $OH(A^2\Sigma^+)$ fragments, detected via their spontaneous fluorescence, are produced in states of high rotational angular momentum.[24-26] When they are generated via *direct* excitation into the $\tilde{B}^1A_1$ state, those in the highest rotational levels are very strongly aligned, indicating fast dissociation. However, the alignment referenced to the *E*-vector of the photolysis beam declines when the less excited rotational levels are monitored, suggesting that dissociation into these channels involves an appreciable delay following photon absorption;[26] more of this later.

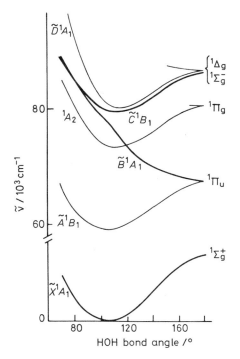

Figure 2 *Bending potentials for the lower electronic states of* H_2O *(*$r_{O—H} = 0.96$ Å*)* (Adapted from *Mol. Phys.*, 1986, **57**, 129)

[24] T. Carrington, *J. Chem. Phys.*, 1964, **41**, 2012.
[25] J. P. Simons and A. J. Smith, *Chem. Phys. Lett.*, 1983, **97**, 1.
[26] J. P. Simons, A. J. Smith, and R. N. Dixon, *J. Chem. Soc., Faraday Trans. 2*, 1984, **80**, 1489.

The level of understanding summarized so far was achieved entirely on the basis of spectroscopic and photochemical investigations using single-photon absorption techniques in the vacuum ultraviolet (v.u.v.). A new insight into the predissociation of the rotationally structured Rydberg states was provided by the $3 + 1$ REMPI experiments of Dixon, Ashfold, and their co-workers[27] (see Figure 1). Taking the transition $\tilde{C}^1B_1 \leftarrow \tilde{X}^1A_1$ as an illustration, coherent 3-photon absorption of tunable dye laser radiation at 372 nm populates selected rotational levels $J_{K_aK_c}$ in the (0,0,0) vibrational level of the $\tilde{C}$ state. Their subsequent fate depends on the competition between predissociation and secondary incoherent absorption of a further photon transferring the excited molecule into the ionization continuum. The widths and intensities of rotational features in the REMPI spectrum become broader and weaker as the rates of predissociation increase. The spectral analysis enabled the predissociation rates to be determined absolutely, and also established the operation of both a homogeneous and a heterogeneous predissociation pathway. The latter was promoted by rotation about the a-inertial axis, consistent with predissociation via $\tilde{B}^1A_1$; the former presumably involves the dissociative $\tilde{A}^1B_1$ state.

If the photoexcited $H_2O(\tilde{C}^1B_1)$ or $D_2O(\tilde{C}^1B_1)$ can live long enough to absorb a further photon and ionize, there must also be the possibility that it might be able to emit a photon spontaneously and to decay radiatively either to the ground state or to the $\tilde{A}^1B_1$ continuum. Of the two alternatives the latter would be the more favoured, since it would be associated with the transition $(1b_1)^1(3pa_1)^1 \rightarrow (1b_1)^1(3sa_1)^1$, the $3p \rightarrow 3s$ analogue of the Na D-line! The transition to the ground state, $(1b_1)^1(3pa_1)^1 \rightarrow (1b_1)^2$, would be forbidden by the Laporte selection rule to the extent that the orbital $1b_1$ could be described as $2p$ centred on the O atom. This prediction was confirmed by the detection of a broad, bound-to-free fluorescence continuum centred in the violet region of the spectrum, following two-photon excitation of H_2O or D_2O by KrF laser radiation at 248 nm,[28] which, by great serendipity, is 2×124 nm. The excitation spectrum of this fluorescence, generated by scanning a narrow-line injection-locked KrF laser across (the accessible portion of) the $\tilde{C} \leftarrow \tilde{X}$ band, confirmed the assignment of the fluorescence, and its analysis reinforced the quantitative results obtained from the $3 + 1$ REMPI experiment. The $3p \rightarrow 3s$ Rydberg molecular fluorescence is by no means unique: it is also found in the $\tilde{C}' \rightarrow \tilde{A}$ emission of NH_3 and ND_3,[29] where, because the $\tilde{C}'$ state is only very weakly predissociated, it has a quantum yield high enough to support laser action;[30] additionally it gives rise to the Schüler emission bands that have been

[27] M. N. R. Ashfold, *Mol. Phys.*, 1986, **58**, 1.

[28] M. P. Docker, A. Hodgson, and J. P. Simons, *Mol. Phys.*, 1986, **57**, 129.

[29] M. N. R. Ashfold, C. L. Bennett, R. N. Dixon, P. Fielden, H. Rieley, and R. J. Stickland, *J. Mol. Spectrosc.*, 1986, **117**, 216.

[30] C. R. Quick, jun., J. H. Glownia, J. J. Tiee, and F. L. Archuleta, Proc. 1983 Los Alamos Conf. Opt., SPIE, 1983, p. 156.

assigned to the NH_4 radical.[31] No doubt many more polyatomic molecular Rydberg emission spectra are waiting to be discovered.

Two-photon absorption at 248 nm also excites a much stronger fluorescence associated with the primary photofragments $OH(A^2\Sigma^+)$, with the majority of fragments populating high rotational levels.[28,32] However, the excitation spectrum of the $OH(A \rightarrow X)$ fluorescence in no way resembles that of the parent molecular fluorescence, since all features with $<J_a^2> = 0$ are absent.[32] The electronically excited fragments $OH(A^2\Sigma^+)$ and $OD(A^2\Sigma^+)$ are products of the heterogeneous predissociation pathway only; the homogeneous pathway leads to 'dark' products, presumably $OH(X^2\Pi)$, as would be expected if this channel were accessed via the $\tilde{A}^1B_1$ continuum directly. The photofragment fluorescence excitation also appears to include a contribution from a broad underlying continuum, upon which the structured $\tilde{C} \leftarrow \tilde{X}$ band is superposed.[32] In contrast, the *parent* molecular fluorescence excitation spectrum was composed of the structured features only.[28] This could be understood if the continuum were associated with direct photodissociation from the underlying $\tilde{B}^1A_1$ state, which also contributes to the one-photon (and presumably two-photon) absorption at 124 nm (248 nm).

These suggestions were confirmed through measurements of the $OH(A)$ and $OD(A)$ photofragment fluorescence polarization[32] recorded for both Q- and P-branch lines in the resolved $OH(A \rightarrow X)$ and $OD(A \rightarrow X)$ spectrum. The corresponding photofragment alignments, $<P_2(\hat{j}\cdot\hat{\varepsilon})>$, increased in proportion to the contribution made by the continuum. The $OH(A)$ and $OD(A)$ fragments generated by excitation into a strong rotational feature in the $\tilde{C}^1B_1$ state showed negligible alignment. Since predissociation from the $\tilde{C}$ state requires out-of-plane rotation, about the a-inertial axis, the loss of alignment is readily understood, similarly the strong alignment following direct dissociation from the underlying $\tilde{B}$-state continuum. Finally, the sign and the absolute magnitude of the maximum recorded alignments, $\mathscr{A}_0^{(2)} > + 0.6$, confirmed the upper-state(s) symmetry as predominantly A_1 rather than the possible alternative A_2, arising from the neighbouring $^1A_2$ state, accessible in principle by two-photon absorption (see Figure 2).*

Branching via the $\tilde{B}^1A_1$ Potential-Energy Surface. The importance of rotation in controlling the branching into alternative electronic channels was further amplified by detailed analysis of the photofragment fluorescence excitation spectra. The intensities of contributing features could only be reproduced if the yields of $OH(A)$ and $OD(A)$ fragments passed through a maximum as the expectation value $<J_a^2>$ of the rotational angular momentum increased.[28,32] This behaviour reflects the dynamics on the $\tilde{B}^1A_1$ potential following the radiationless transfer $\tilde{C} \rightarrow \tilde{B}$. As the bond angle

---

* Absorption into the $^1A_2$ state could make a contribution to the overall dissociation process, of course, through the alternative *un*observed channel, e.g. $H_2O + h\nu \rightarrow H_2 + O(^1D)$[34,35]

[31] J. K. G. Watson, *J. Mol. Spectrosc.*, 1984, **107**, 124.
[32] A. Hodgson, J. P. Simons, M. N. R. Ashfold, J. M. Bayley, and R. N. Dixon, *Chem. Phys. Lett.*, 1984, **107**, 1; *Mol. Phys.*, 1985, **54**, 351.

opens and the nuclear configuration approaches linearity, a second transfer can occur, $\tilde{B}^1 A_1 \rightarrow \tilde{A}^1 B_1$, which, like the first, is also promoted by the Coriolis coupling associated with rotation about the a-inertial axis.[33] Since dissociation from the $\tilde{A}$ state generates electronically unexcited fragments OH(X) and OD(X), the fluorescence yield falls off at $<J_a^2> \gg 0$.

Further subtleties were revealed when the fluorescence excitation spectra of the OH(A) and OD(A) fragments in individual selected rotational levels were recorded. The quantum yields of fragments populating the lowest occupied levels fell rapidly with increasing $<J_a^2>$, but for the fragments in the highest levels the dependence on $<J_a^2>$ was much weaker. The higher the value of $<J_a^2>$, the shorter the lifetime in the $\tilde{B}$ state (because of the rotationally dependent $\tilde{B} \rightarrow \tilde{A}$ transfer). Thus the most highly rotating fragments must be associated with the shortest-duration trajectories, while the less excited fragments must be generated through longer-lived trajectories, able to survive several excursions across the Renner–Teller intersection near linearity.

A correlation between high rotational excitation and direct trajectories and between lower excitation and angle-bending 'resonances' had been predicted by Segev and Shapiro[36] on the basis of three-dimensional quantum-scattering calculations. These assumed the *ab initio* potential of Flouquet and Horsley,[37] which presents a weak radial ridge associated with the avoided crossing of $\tilde{B}^1 A_1$ and $\tilde{D}^1 A_1$ (though more recent calculations suggest that the ridge may actually be less evident than was thought[38]). The correlation had also been suggested by the rotationally resolved photofragment alignment measurements discussed earlier.[25,26] Figure 3a shows a schematic representation of the potential surface and trajectories that would lead to alternative dissociation dynamics. The central minimum corresponds to the conical intersection of the $\tilde{B}^1 A_1$ and ground-state $\tilde{X}^1 A_1$ potentials. The alternative representation of the intersecting potentials, Figure 3b, shows that the trajectory '0' which passes through linearity on the outer wall of the conical intersection avoids the Renner–Teller region. It leads to OH(A) and OD(A) in high rotational levels. The trajectories '1' and '2' passing inside the conical intersection cross the Renner–Teller region once and twice, respectively. Leakage into the $\tilde{A}$ state increases with the number of crossings, reducing the yield of OH(A) and OD(A) fragments and the rotational excitation in those that are produced. Segev and Shapiro predicted some trajectories that were particularly long-lived and would lead to sharp 'resonances' in the $\tilde{B}^1 A_1 \leftarrow \tilde{X}^1 A_1$ photodissociation continuum. To date, no such resonance structures have been detected. The rapid leakage $\tilde{B} \rightarrow \tilde{A}$ and/ or $\tilde{B} \rightarrow \tilde{X}$, not included in the quantum calculations, would of course

[33] R. N. Dixon, *Mol. Phys.*, 1985, **57**, 333.
[34] S. Tsurubuchi, *Chem. Phys.*, 1975, **10**, 335.
[35] C. R. Claydon, G. A. Segal, and H. S. Taylor, *J. Chem. Phys.*, 1971, **54**, 3799.
[36] E. Segev and M. Shapiro, *J. Chem. Phys.*, 1982, **77**, 5604.
[37] F. Flouquet and J. A. Horsley, *J. Chem. Phys.*, 1974, **60**, 3767.
[38] G. Theodorakopoulos, I. Petsalakis, and R. J. Buenker, *Chem. Phys.*, 1985, **96**, 217.

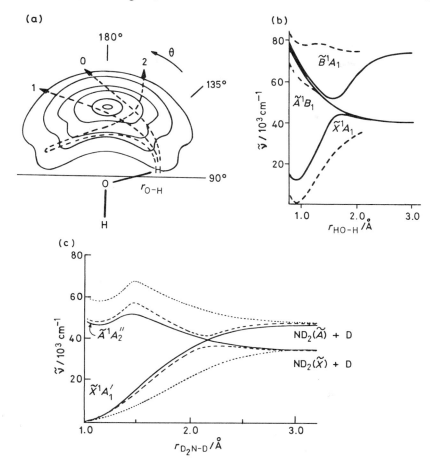

Figure 3 (a) *Schematic potential-energy surface for the $\tilde{B}^1A_1$ state of H_2O showing alternative types of dissociative trajectory. 0: direct, passing outside both the Renner–Teller and the conical intersections; 1 and 2: trajectories that pass inside the conical i.tersection and cross the Renner–Teller intersections once and twice. (b) Asymmetric stretching potentials for the ground and lower-lying electronic states of H_2O; one of the bond lengths is fixed at 0.96 Å. The solid curves are potentials for bond angles near linearity ($\theta = 170°$) and the dotted curves are for $\theta = 104.5°$. (c) Schematic stretching potentials for the lower-lying electronic states of ND_3. Solid curves for out-of-plane angle $\theta = 0°$, dashed curve $\theta \simeq 5°$, dotted curve $\theta \simeq 30°$*

(Reproduced, in part, with permission from *Mol. Phys.*, 1986, **57**, 129)

suppress them, but a recording of the photofragment fluorescence excitation spectrum under jet-cooled conditions, where $<J_a^2> \sim 0$ and the $\tilde{B} \to \tilde{A}$ transfer would be strongly suppressed, still shows no indication of resonance structure.[39] If the predicted sharp resonances *are* in fact real and not an

[39] M. Brouard, M. P. Docker, A. Hodgson, and J. P. Simons, *Faraday Discuss. Chem. Soc.*, 1986, **82**, in press.

artefact arising from either the simplifying assumptions employed in the scattering calculations or the potential surfaces used, their absence at low temperature would appear to indicate an important contribution from the $\tilde{B} \to \tilde{X}$ leakage path. Like the alternative $\tilde{B} \to \tilde{A}$ path, this would lead to the formation of OH(X) and OD(X) fragments. Indeed, the average quantum yields of the *un*excited OH(X) fragments are $>90\%$.[40-42] The lack of success in monitoring their production directly, through LIF detection, has been attributed to their production in high rotational (and/or vibrational) levels, for which u.v. laser excitation leads not to fluorescence but to predissociation.

A recent, and very elegant, experiment by Welge and his co-workers[43] has confirmed this assumption. Measurements of the velocity spectrum of the recoiling H atoms established an internal-energy distribution in the partner OH fragments consistent with the predominant production of OH(X) in rotational levels $N \sim 30$—48, but with no vibrational excitation. However, arguments were also advanced to suggest that the $\tilde{B} \to \tilde{X}$ pathway was *un*important compared with $\tilde{B} \to \tilde{A}$. Finally, to add to the confusion, calculations of non-adiabatic transfer probabilities from $H_2O(\tilde{B})$ by Dunne and Murrell[44] conclude that $\tilde{B} \to \tilde{X}$ transfer would lead to OH(X) fragments carrying high vibration rather than rotation. Since this conflicts, apparently, with the analysis of the time-of-flight velocity spectra, one might conclude that the $\tilde{B} \to \tilde{X}$ pathway can be downgraded after all. Sufficient to say that the relative importance of the alternative electronic branching paths remains uncertain.

The story that has been unfolded reviews a wealth of detail in the photodissociation dynamics of H_2O and D_2O from the $\tilde{C}$ and $\tilde{B}$ states. Much additional information has been revealed for the more highly excited Rydberg states using REMPI techniques by Ashfold, Dixon, and their co-workers.[27] Taken together with the penetrating and ingenious experimental studies of $H_2O(\tilde{A}^1B_1)$ by Andresen and his co-workers[45,46] (see Chapter 3) and the complementary theoretical analysis by Schinke, Balint-Kurti, and others,[47] we suspect that more is now known about this model photodissociation system than about any other. Some things are still obscure of course, particularly the dynamics of the 'dark' channels leading to OH($X^2\Pi$) and OD($X^2\Pi$) or to H_2 and $D_2 + O(^1D, ^3P)$. The $\tilde{D}''^1A_2$ Rydberg state, detected via REMPI spectroscopy, is heavily predissociated by a neighbouring A_2

[40] I. P. Vinogradov and F. I. Vilesov, *Opt. Spectrosc.*, 1976, **40**, 32.

[41] L. C. Lee, *J. Chem. Phys.*, 1980, **72**, 4334.

[42] O. Dutuit, A. Tabche-Fouhaile, I. Nenner, H. Frohlich, and P. M. Guyon, *J. Chem. Phys.*, 1985, **83**, 584.

[43] H. J. Krautwald, L. Schnieder, K. H. Welge, and M. N. R. Ashfold, *Faraday Discuss. Chem. Soc.*, 1986, **82**, in press.

[44] L. J. Dunne and J. N. Murrell, *Faraday Discuss. Chem. Soc.*, 1986, **82**, in press.

[45] P. Andresen, G. S. Ondrey, B. Titze, and E. W. Rothe, *J. Chem. Phys.*, 1984, **80**, 2548.

[46] P. Andresen, V. Beushausen, D. Häusler, H. W. Lülf, and E. W. Rothe, *J. Chem. Phys.*, 1985, **83**, 1429.

[47] R. Schinke, V. Engel, P. Andresen, D. Häusler, and G. G. Balint-Kurti, *Phys. Rev. Lett.*, 1985, **55**, 1180.

continuum, presumably the lowest $^1A_2$ state[48] (see Figure 2), and both the continuum and its dissociation products [thought to be $H_2 + O(^1D)$][34,35] remain experimentally uncharacterized. There is still work to be done.

NH_3 *and* ND_3

In NH_3 the addition of a third hydrogen atom binds one of the two lone pairs of electrons previously involved in the lower-lying electronic transitions in water, and only the totally symmetric non-bonding orbital $3a_1$ (C_{3v}) is left. The first electronic transition in ammonia involves the promotion $3sa_1 \leftarrow 3a_1$: it populates a planar excited state, $\tilde{A}^1A_2''$ (D_{3h}), and is the analogue of the bent-to-linear transition $\tilde{B}^1A_1 \leftarrow \tilde{X}^1A_1$ in water. Since the erstwhile lone-pair $1b_1$ orbital in water is now a bonding orbital, there is no analogue for the first continuum, $\tilde{A}^1B_1 \leftarrow \tilde{X}^1A_1$. The schematic potentials are compared in Figure 3. Once again the ground-state potential correlates with the electronically excited molecular fragment [$NH_2(\tilde{A}^2A_1)$], and there is a conical intersection involving $NH_3(\tilde{A})$ and $NH_3(\tilde{X})$ (reminiscent of the $\tilde{B}$–$\tilde{X}$ intersection in H_2O). Under C_s symmetry the $NH_3(\tilde{A})$ and $NH_3(\tilde{X})$ states both become A'. As the H_2N—H separation increases, the molecular potential passes over a pronounced barrier, associated with the evolution of the upper molecular-orbital character from Rydberg, $3sa_1$, to intravalence $4a_1$–σ^* and ultimately to pure $H(1s)$.[49] The barrier height is sufficient to impose a tunnelling mechanism on the predissociation dynamics from the lowest vibrational levels in $NH_3(\tilde{A})$ and $ND_3(\tilde{A})$.[17] This gives the upper state a lifetime long enough to support resolved vibronic structure associated with the umbrella motions, v_2', and, because of the D-isotope effect, some poorly resolved rotational structure in ND_3, particularly in the level 2^1.[50,51] The expected activity in the symmetric stretching mode v_1' cannot be resolved in absorption because of rapid predissociation and perhaps coincidental overlap of v_1' by $3v_2'$.[52,53] However, it can be inferred from the progressions in v_1'' excited in the weak fluorescence emission spectrum of the $\tilde{A}$ state,[17,54] and it can be identified by an experiment that populates vibronic levels in $ND_3(\tilde{A})$ not by absorption but via spontaneous fluorescence from the much longer-lived Rydberg state $ND_3(\tilde{C}')$. The dispersed fluorescence shows a marked asymmetry in the homogeneously broadened spectral profile of levels with $v_1' = 1$, which has been attributed to the interaction of v_1' with the antisymmetric stretching continuum 'nv_3'' lying above the exit barrier on the $\tilde{A}$-state potential.[17]

The resolved rotational structure detectable in the $\tilde{A} \leftarrow \tilde{X}$ absorption system of ND_3 is remarkable for appearing more strongly in the 2_0^1 than the

48 M. N. R. Ashfold, J. M. Bayley, and R. N. Dixon, *Can. J. Phys.*, 1984, **62**, 1806.
49 R. Runau, S. D. Peyerimhoff, and R. J. Buenker, *J. Mol. Spectrosc.*, 1977, **68**, 253.
50 A. E. Douglas, *Faraday Discuss. Chem. Soc.*, 1963, **35**, 158.
51 M. N. R. Ashfold, C. L. Bennett, and R. N. Dixon, *Chem. Phys.*, 1985, **93**, 293.
52 P. Avouris, A. R. Rossi, and A. C. Albrecht, *J. Chem. Phys.*, 1981, **74**, 5516.
53 W. R. Harshbarger, *J. Chem. Phys.*, 1970, **53**, 903.
54 L. D. Ziegler, P. B. Kelly, and B. Hudson, *J. Chem. Phys.*, 1984, **81**, 6399.

2^0_0 transitions.[50,51] This vibronic-state dependence has been explained by assuming a potential barrier to predissociation that is at a minimum and hence most easily penetrated in a planar configuration.[17,51] The eigenfunction associated with the bending motion has a zero amplitude for planar geometry when $v_2' = 1$ but not when $v_2' = 0$ or 2. An additional rotational-state dependence for the predissociation rate has been revealed by analysis of the rotationally resolved two-photon LIF excitation spectrum of $ND_3(\tilde{A})$, 2^1_0. The excitation spectrum is well reproduced by assuming a damping function of the form:[51]

$$f(J,K) = 1 + a[J(J + 1) - bK^2] \qquad (1)$$

and a qualitative analysis concludes that predissociation is promoted by the centrifugal distortion associated with rotation about the (in-plane) b-inertial axis.[51] When $v_2' > 1$, however, the rotational dependence can no longer be discerned.[17]

The products of predissociation from $NH_3(\tilde{A})$ and $ND_3(\tilde{A})$ are exclusively NH_2 or $ND_2 + H$ or D, with all but a few per cent of the molecular fragments in the ground electronic state.[15,55] Those that are formed in the excited state, $NH_2(\tilde{A}^2A_1)$ and $ND_2(\tilde{A}^2A_1)$, arise from the small fraction of parent molecules that are not 'caught' at the conical intersection.[51] The excited fragments generated by laser photodissociation of NH_3 at 193 nm[55] are rotationally aligned, with a strong preference for rotation about their a-inertial axis; this suggests that the small fraction corresponds to those molecules that penetrate the predissociation barrier whilst still in a non-planar configuration. Out-of-plane motion in the parent molecule would evolve into the observed photofragment rotational alignment.

Resolution of the severely broadened short-lived rovibronic levels in $NH_3(\tilde{A})$ (as opposed to ND_3) has had to await the advent of the rotationally selective two-colour double-resonance techniques, based on the much longer-lived $\tilde{C}'^1A'_1$ state.[16,17] This can be accessed via $3hv$[16] or $2hv$[29] excitation and depopulated by further photon absorption into the ionization continuum[16] or by stimulated emission pumping into single rovibronic levels in the $\tilde{A}$ state[17] (or by spontaneous fluorescence or weak predissociation[29]). Variations in the ion current[16] or the spontaneous-fluorescence intensity[17] as the second laser is scanned – the ion or fluorescence 'dip' – trace the resolved or partially resolved *single* rotational features in the $\tilde{C}' \rightarrow \tilde{A}$ spectrum. Since the low-lying rovibronic level widths in the $\tilde{C}'$ state are orders of magnitude narrower than those in the $\tilde{A}$ state, corresponding in ND_3 to lifetimes approaching the nanosecond range,[29] the homogeneous widths of rotational features in the double-resonance spectrum allow the determination of predissociation rates from quantum-state-selected levels in the $\tilde{A}$ state (provided care is taken to minimize and, hopefully, to eliminate any distortions arising from power broadening). Table 1 summarizes current data for $NH_3(\tilde{A})$ and $ND_3(\tilde{A})$.

[55] V. M. Donnelly, A. P. Baronavski, and J. R. McDonald, *Chem. Phys.*, 1979, **43**, 271.

Table 1 *Bandwidths and predissociation rates in selected vibronic levels of*
$NH_3(\tilde{A}^1A_2'')$ *and* $ND_3(\tilde{A}^1A_2'')$ *(taken from ref. 17)*

	Vibronic level	$FWHM/cm^{-1}$	τ/ps
NH_3	0^0	41 ± 5	0.13 ± 0.02
	2^1	33 ± 5	0.16 ± 0.02
	2^2	43 ± 5	0.12 ± 0.02
	$1^1 2^2$	~ 500	~ 0.01
ND_3	0^0	4.5	1.2
	2^1	1.1	4.8
	2^2	23 ± 4	0.23 ± 0.04
	2^3	35 ± 10	0.15 ± 0.04
	$1^1 2^2$	$\leqslant 250$	$\geqslant 0.02$

These double-resonance experiments represent (one of) the state-of-the-art approaches to the study of quantum-state-selected predissociation. However, the understanding of the photodissociation dynamics in NH_3 and ND_3 is still not quite as developed as in H_2O and D_2O, principally because of the absence of parallel-state selection in the photofragments. That is still to come.

3 Photofragment Vector Correlations

The absorption of a photon necessarily prepares a population of excited molecules that are aligned in the laboratory frame, with their transition dipoles $\mathbf{\mu}$ preferentially directed parallel to the ε-vector of the absorbed photon, ε_p.[4,5,56] When excitation is followed by molecular photodissociation the anisotropy may be carried over into the photofragments, but the observed anisotropies will also depend upon the dynamics of the photodissociation process and the conservation laws. They are measured by the vector correlations introduced briefly in Section 1, between the molecular transition moment, $\mathbf{\mu}$, the centre-of-mass recoil velocity, $\mathbf{v}$, and the rotational angular momentum of the fragment(s), $\mathbf{j}$. Provided dissociation is fast enough to preserve the correlation between ε_p (laboratory frame) and $\mathbf{\mu}$ (molecule frame), *i.e.* fast compared to the excited molecular rotational period, the correlations with $\mathbf{\mu}$ can be inferred from laboratory measurements of angle-resolved photofragment velocity distributions and rotational alignments, $\mathscr{A}_0^{(2)}$.

(v, μ)

The first of these to be examined was the angular scattering distribution, (v, μ), first pioneered in bulb deposition experiments[57-59] and in time-of-flight

[56] R. Bersohn and S. H. Lin, *Adv. Chem. Phys.*, 1969, **16**, 67.
[57] J. Solomon, *J. Chem. Phys.*, 1967, **47**, 889.
[58] C. Jonah, P. Chandra, and R. Bersohn, *J. Chem. Phys.*, 1971, **55**, 1903.
[59] J. Solomon, C. Jonah, P. Chandra, and R. Bersohn, *J. Chem. Phys.*, 1971, **55**, 1908.

(TOF) angular resolved scattering measurements by Wilson[2,60-63] and co-workers. The angular distribution function for the recoiling fragments takes the form:

$$I(\theta) = \frac{1}{4\pi} [1 + \beta P_2(\cos \theta)] \tag{2}$$

where $\cos \theta = \hat{v} \cdot \hat{\varepsilon}_p$ and β is the anisotropy parameter. Its value reflects the average angle between the recoil axis v and the transition moment μ. For fast dissociation $\beta = +2 \, (v \parallel \mu)$ or $-1 \, (v \perp \mu)$; otherwise $2 > \beta > -1$. Thus measurements of the angular distribution can lead via β to assignments of the symmetry of the electronically excited state(s) of the parent molecule and to estimates of its dissociation rate, measured against a rotational 'clock'.[60,63-65]

The TOF distribution, after laboratory → centre-of-mass transformation, gives the kinetic-energy release among the photofragments and, through energy conservation, information on the partitioning of internal energy among the fragments. Indeed, except where optical spectroscopic probing of the recoiling fragments is practical, TOF spectroscopy provides the only source of such information. However, when the fragments include atomic or fluorescent diatomic species, Doppler spectroscopy provides an increasingly attractive alternative. In the simplest case, where the parent molecule is diatomic, observation of the Doppler profile of the atomic fragments via resonance fluorescence[66,67] readily reveals the branching into alternative electronic channels, for example[68] in the dissociation:

$$HI + h\nu \rightarrow H + I(^2P_{\frac{3}{2}}) \text{ or } I(^2P_{\frac{1}{2}}) \tag{3}$$

monitored via the Lyman-α Doppler profile of the H atom. When the parent molecule is triatomic or polyatomic, the distribution over internal states in the molecular fragment(s) generally creates a broad distribution of recoil velocities, and the resolution of the Doppler profiles into angle and velocity distributions requires probing at more than one geometry.[69] However, if one of the products is monoatomic, as is inevitable for a triatomic molecule such as ICN, Doppler spectroscopy of selected vibrational–rotational states in the *diatomic* fragment, using narrow-line LIF probing, necessarily recovers the Doppler resolution that obtains when the parent is diatomic. Using this

[60] G. E. Busch, R. T. Mahoney, R. I. Morse, and K. R. Wilson, *J. Chem. Phys.*, 1969, **53**, 449, 837.
[61] R. J. Oldman, R. K. Sander, and K. R. Wilson, *J. Chem. Phys.*, 1971, **54**, 4127.
[62] G. E. Busch and K. R. Wilson, *J. Chem. Phys.*, 1972, **56**, 3626, 3638.
[63] J. H. Ling and K. R. Wilson, *J. Chem. Phys.*, 1975, **63**, 101.
[64] C. Jonah, *J. Chem. Phys.*, 1971, **55**, 1915.
[65] S.-C. Yang and R. Bersohn, *J. Chem. Phys.*, 1974, **61**, 4400.
[66] R. N. Zare and D. R. Herschbach, *Proc. IEEE*, 1963, **51**, 173.
[67] R. N. Zare, *Mol. Photochem.*, 1972, **4**, 1.
[68] R. Schmiedl, H. Dugan, W. Meier, and K. H. Welge, *Z. Phys.*, 1982, **304A**, 137.
[69] J. L. Kinsey, *J. Chem. Phys.*, 1977, **66**, 2560; *Annu. Rev. Phys. Chem.*, 1977, **28**, 349.

approach Wittig and co-workers[70] were able to separate the rotational population distributions in the CN(X) fragments associated with $I(^2P_{\frac{3}{2}})$ and $I(^2P_{\frac{1}{2}})$ and to determine the anisotropy parameters, β, for each channel. In another approach Wittig's group have demonstrated an ingenious alternative means of simplifying the experimental data. The introduction of sufficient delay between the laser photolysis pulse and the subsequent laser probe[71] allows all fragments, save those recoiling along the probe beam axis, to disappear out of sight, so that the double-peaked 'forward–backward' Doppler profile directly reflects the velocity distribution along the chosen axis.

(j, μ)

The second correlation, (j, μ), between the rotational angular momentum of a diatomic fragment and the parent molecular transition moment, has been studied in a wide range of systems both through measurements of spontaneous-fluorescence polarization and through the polarization dependence of LIF.[1] The theory for each type of measurement, applied specifically to the alignment/polarization of photodissociation products, has been presented by Greene and Zare[4,5] to provide a clear guide for the experimentalist. With the polarization vector ε_p of the absorbed photon defining the laboratory z-axis, at the high-j limit the maximum observed alignments $\mathscr{A}^{(2)}_0 = 2 < P_2(\hat{j} \cdot \hat{z}) >$ may take the values $+\frac{4}{5}$ ($j \parallel \mu$) or $-\frac{2}{5}$ ($j \perp \mu$). In practice the observed alignments may fall short of these limits for a number of reasons. These may include (i) the photodissociation dynamics, (ii) precessional motion associated with coupling of the fragment nuclear rotation to internal (electron and/or nuclear spin) or external (*e.g.* Earth's) magnetic fields,[4,5,26] and (iii) rotational motion in the photoexcited molecule prior to dissociation[72] (*cf.* the reduction in the magnitude of the asymmetry parameter β).

Finally, inherent in the theoretical analysis is the assumption that the measurements of photofragment polarization are averaged over all fragment recoil velocities, *i.e.* that the probe laser linewidth is much greater than the Doppler width due to fragment recoil. In practice this condition may frequently *not* be met when LIF probing is employed,[73,74] and in this case polarization measurements require averaging across the Doppler-broadened spectral-line profile.

(v, j) *and Beyond*

There is a rapidly growing range of systems in which the energy and angular momentum disposal in the diatomic products of photodissociation has been

[70] I. Nadler, D. Mahgerefteh, H. Reisler, and C. Wittig, *J. Chem. Phys.*, 1985, **82**, 3885.

[71] Z. Xu, B. Koplitz, S. Buelow, D. Baugh, and C. Wittig, *Chem. Phys. Lett.*, 1986, **127**, 534.

[72] T. Nagata, T. Kondow, K. Kuchitsu, G. W. Loge, and R. N. Zare, *Mol. Phys.*, 1983, **50**, 49.

[73] J. A. Guest, M. A. O'Halloran, and R. N. Zare, *Chem. Phys. Lett.*, 1984, **103**, 261.

[74] H. Joswig, M. A. O'Halloran, R. N. Zare, and M. S. Child, *Faraday Discuss. Chem. Soc.*, 1986, **82**, in press.

probed using LIF techniques. These include $OH(X^2\Pi)$ from $HONO$[75] and H_2O_2,[10−13] $NO(X^2\Pi)$ from $(CH_3)_2NNO$,[14] $CN(X^2\Sigma^+)$ from ICN,[70] and $CO(X^1\Sigma^+)$ from OCS and $(HCO)_2$.[8,9] In each case the analysing probe laser had a linewidth sufficiently narrow to resolve the Doppler shift associated with the fragment recoil (typically $\Delta\nu_D \sim 0.5\,cm^{-1}$), and the geometry was generally set with the propagation axis $\boldsymbol{k}_a \parallel \boldsymbol{\varepsilon}_p$ and/or $\boldsymbol{k}_a \perp \boldsymbol{\varepsilon}_p$. In cases where the second fragment was monoatomic, *e.g.* ICN and OCS, the detected fragment can be taken to have a single centre of mass recoil velocity (or velocities, where there are alternative spin–orbit channels). In the general case there may be a broader spread of velocities, but in favourable systems (*e.g.* H_2O_2) the assumption of a single recoil velocity may still be a good approximation; this greatly simplifies the analysis of the Doppler profiles.[7,10,11,14] Selection of a particular Doppler shift $\Delta\nu$, by the tunable probe laser, thus allows selection of a subset of the fragments associated with a particular scattering direction. When $\boldsymbol{j}$ and $\boldsymbol{v}$ are correlated, as will generally be true,[7−14,75,76] the alignment of this subset will also change with the selected Doppler shift. As a result, the observed Doppler profiles will be sensitive not only to the choice of analysis probe directions, $\boldsymbol{k}_a \parallel \boldsymbol{\varepsilon}_p$ or $\boldsymbol{k}_a \perp \boldsymbol{\varepsilon}_p$, but also to the choice of polarization, $\boldsymbol{\varepsilon}_a \parallel \boldsymbol{\varepsilon}_p$ or $\boldsymbol{\varepsilon}_a \perp \boldsymbol{\varepsilon}_p$, and to the rotational branch probed, $\Delta J = 0$ or ± 1 (which would otherwise be equivalent). Under these conditions reliable estimates of the translational anisotropy parameters, β, must accommodate the polarization dependence of the measured Doppler profiles. More importantly, the polarization dependence contains novel information on the third vector correlation $(\boldsymbol{v}, \boldsymbol{j})$. As with the other two correlations, two limiting cases can be identified: $\boldsymbol{j} \perp \boldsymbol{v}$, as in the case of a triatomic molecule fragmenting through a bent configuration, or $\boldsymbol{j} \parallel \boldsymbol{v}$, resulting from a torsional motion about the dissociation axis.

Dixon[7] has provided a theoretical description of the correlation between $\boldsymbol{v}$ and $\boldsymbol{j}$ for a molecular fragment, allowing the experimental data to be reduced to a set of bipolar moments, $\beta_0^K(k_1k_2)$, which describe the translational and rotational angular distributions. The spectral lineshapes for LIF detection may be written:

$$g(\bar{\nu}) = \frac{1}{2\Delta\bar{\nu}_D}[g_0 + g_2P_2(\chi_D) + g_4P_4(\chi_D) + g_6P_6(\chi_D)] \qquad (4)$$

where $\chi_D = \Delta\bar{\nu}/\Delta\bar{\nu}_D$ and the terms g_{k_1} are linear combinations of the bipolar moments $\beta_0^K(k_1k_2)$ with coefficients that depend on the choice of beam geometry and spectral-branch polarization.[7] In practice the higher terms P_4 and P_6 are unlikely to be resolved experimentally (since they have a finer structure that is rapidly averaged by the experimental broadening). The rotational alignment, $\mathscr{A}_0^{(2)} = \frac{4}{5}\beta_0^2(02)$, only enters the leading term g_0 and is therefore most conveniently measured separately in the normal way (after ensuring that the *entire* line profile has been integrated). This leaves three

[75] R. Vasudev, R. N. Zare, and R. N. Dixon, *Chem. Phys. Lett.*, 1983, **96**, 399; *J. Chem. Phys.*, 1984, **80**, 4863.
[76] D. A. Case, G. M. McClelland, and D. R. Herschbach, *Mol. Phys.*, 1978, **35**, 541.

remaining moments, which together describe the observed Doppler line shape:* (i) $\beta_0^2(20)$, equal to $\frac{1}{2}\beta$, the conventional translational anisotropy parameter, (ii) $\beta_0^0(22)$, equal to $<P_2(\hat{v}\cdot\hat{j})>$, an isotropic term that is a measure of the mutual correlation between v and j, and (iii) $\beta_0^2(22)$, describing the mutual correlation of v and j with the parent absorption moment μ. These bipolar moments can be determined from the experimental lineshapes recorded for different probe/photolysis polarization geometries and different rotational branches in the LIF excitation spectrum, to provide a means of parameterizing the vector correlations free from any assumed model for the dissociation process.[7] In particular, the isotropic character of the term $\beta_0^0(22)$ implies that the correlation (v, j) may still be observed *even after rotation of the parent molecule has destroyed all memory of the initial photoselection in the LAB frame.* This contrasts with the more common measurements of $\mathscr{A}_0^{(2)}$ [or $\beta_0^2(02)$] or β [or $\beta_0^2(20)$], which are susceptible to memory loss when dissociation is delayed. Measurements of the correlation (v, j) [via $\beta_0^0(22)$] therefore provide a new source of dynamical information reflecting interactions in the exit channel.

Finally, two caveats: the neglect of terms higher than P_2 in equation (4) may not always be justified, and it may lead to some bias in the analysis. More importantly, the assumption of a single recoil velocity will be unrealistic in many (most) cases. Even if simplifying assumptions about the velocity and angular distributions, $W(v)$ and $\beta(v)$, respectively, can be made, *e.g.* the assumption of a statistical distribution for long delay times in polyatomic systems, it will be unlikely that anything beyond the sign of $\beta_0^0(22)$ can be extracted.

Examples from Life

The advent of Doppler spectroscopic probing techniques has led, in a very short time, to a range of experimental studies that illustrate a full range of possible vector correlations in diatomic molecular photofragments. These include the following: OCS,[8,9] where j_{CO} is constrained to lie perpendicular to v; $(CH_3)_2NNO$,[14] where j_{NO} tends to be directed perpendicular to v; H_2O_2,[10-13] where j_{OH} tends to be directed parallel to v; $(HCO)_2$,[9] where a correlation between v and j_{CO} can still be identified despite the very long dissociation time ($\sim 10^{-6}$ s) and the existence of more than one dissociative channel.

OCS. The simplest system studied so far is the dissociation of OCS at 222 nm.[8,9] Conservation of angular momentum requires:

$$j_{OCS} + j_{ph} = j_S + j_{CO} + l \tag{5}$$

---

*A fourth moment, $\beta_0^2(24)$, also contributes, but very weakly because its multiplying coefficient is relatively small[7]

but in practice jet-cooling of the OCS and high rotational excitation of the CO photofragments ensure $j_{CO} \gg j_{OCS}, j_{ph}$, and j_S, so that:

$$j_{CO} \simeq -l \qquad (6)$$

i.e. the orbital angular momentum, l, and the fragment rotation are anti-parallel. l is also necessarily perpendicular to the recoil velocity, v, so that $j_{CO} \perp v$. Not surprisingly, analysis of the Doppler profiles of the *P*-, *R*-, and *Q*-branches of the LIF excitation spectrum of $CO(A \leftarrow X)$ led to the same conclusion. Additionally, application of *energy* conservation to the expected velocity distributions showed that both higher ($j_{CO} = 66$) and lower ($j_{CO} = 56$) rotational levels in CO were associated predominantly with the same spin–orbit channel:[8,9]

$$OCS + hv(222 \text{ nm}) \rightarrow CO(X) + S(^1D) \qquad (7)$$

H_2O_2. The best characterized system in which the vector correlations are non-trivial is provided by the photodissociation:[10-13]

$$H_2O_2 + hv(266 \text{ nm or } 248 \text{ nm}) \rightarrow 2OH(X)_{v=0, N} \qquad (8)$$

which generates $OH(X)$ fragments with no vibrational excitation, modest rotational excitation ($<f_R> = 0.11$ at 248 nm), but high kinetic energy ($v_{OH} \simeq 4 \text{ km s}^{-1}$ at 248 nm). The assumption of a near monoenergetic velocity distribution for $OH(X)$ fragments in any selected rotational level provides a very good, though not quite perfect, match to the observed Doppler shifts.[10,11] A broader distribution would have been a serious handicap to analysis of the Doppler profiles, requiring both numerical deconvolution (to accommodate the probe laser linewidths) *and* numerical differentiation (to establish the translational-energy distribution); it is doubtful whether the signal-to-noise ratio in the experimental data would be adequate to sustain this.

The LIF Doppler profiles of the $OH(X)$ fragments were sensitive to the choice of photolysis probe beam polarization and geometry, to the choice of rotational branch, and to the rotational level, N. Some typical profiles are shown in Figures 4 and 5. The translational anisotropies determined through their analysis were $\beta [= 2\beta_0^2(20)] = -1.0 \pm 0.1$ at 248 nm[10,11] and $\beta = -0.71 \pm 0.08$ at 266 nm.[12,13] The negative sign indicates a preference for $v \perp \mu$, which supports the theoretical assignment[77] of the long-wavelength absorption continuum in H_2O_2, to the transition $\tilde{A}^1A \leftarrow \tilde{X}^1A$ with μ parallel to the C_2 symmetry axis. The lower anisotropy at the longer wavelength was attributed to the possibility of a slightly delayed photodissociation.

The rotational alignments in the most highly excited rotational levels were found to be small, but positive with $\mathscr{A}_0^{(2)} [\equiv \frac{4}{5}\beta_0^2(02)] = 0.11$ (248 nm)[11] and

[77] C. Chevaldonnet, H. Cardy, and A. Dargelos, *Chem. Phys.*, 1986, **102**, 55.

Figure 4 *Doppler profiles of OH($X^2\Pi$) generated via photodissociation of H_2O_2 at 248 nm. Line profile for the R_2 (1) line (for which there is no alignment dependence), with the analysing laser beam wave vector, k_a, perpendicular to the photolysis beam polarization vector, ε_p. The simulation is for an asymmetry parameter $\beta = -1$, the limiting value for perpendicular recoil.*
(Reproduced, with permission from *Faraday Discuss. Chem. Soc.*, 1986, **82**, in press)

0.10 (266 nm).[13] However, a more interesting feature of the Doppler profiles was their indication of a much stronger correlation between the fragment rotational and translational vectors, j_{OH} and v, respectively, with a clear preference for rotation parallel to the recoil axis. This correlation is expressed in the positive value derived for the bipolar moment $\beta_0^0(22) = \langle P_2(\hat{j}_{OH} \cdot \hat{v}) \rangle$, which for high rotational levels and photodissociation at 266 nm takes the value $\beta_0^0(22) = +0.35 \pm 0.04$.[13] The positive correlation between v and j_{OH} can be visualized, qualitatively, by a comparison of the Doppler profiles of the Q- and P,R-branches in the LIF spectra (see Figure 5). For an unpolarized photolysis beam, and coaxial probing ($k_p \parallel k_a$, $\varepsilon_p \perp k_a$), the rotation will tend to lie parallel to the probe beam axis k_a for those fragments that contribute to the wings of the Doppler-broadened line ($v \parallel k_a$) and lie perpendicular to the axis at the line centre ($v \perp k_a$). It follows that in the wings the contributing fragments tend to have $j_{OH} \perp \varepsilon_a$ and have $j_{OH} \parallel \varepsilon_a$ near the line centre. Thus the intensities of Q-branch lines at high j_{OH} tend to be enhanced at the centre and reduced in the wings, P- and R-branches showing the reverse behaviour, in agreement with experimental observation (see Figure 5).

The identification of the 'cartwheel' rotational motion about the fragment recoil axis allows a detailed analysis of the alternative factors that may

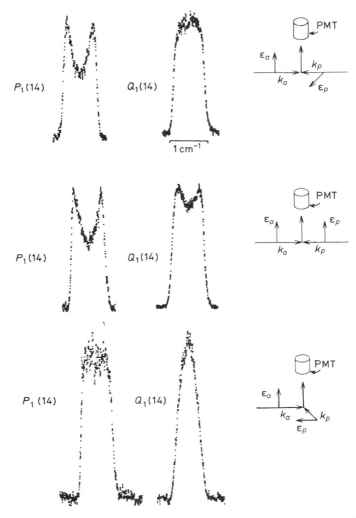

Figure 5 *Doppler profiles of* $OH(X^2\Pi)$ *generated via photodissociation of* H_2O_2 *at 248 nm. Line profiles for* P_1 (14) *and* Q_1 (14) *for three alternative geometries of the analysis* $(\boldsymbol{k}_a, \boldsymbol{\varepsilon}_a)$ *and photolysis* $(\boldsymbol{k}_p \cdot \boldsymbol{\varepsilon}_p)$ *laser beams. Note the changes in contour between the P- and Q-lines, caused by correlations between the recoil velocity,* $\boldsymbol{v}$, *and rotation vector,* $\boldsymbol{J}$. *The contours can be understood if there is a preference for* $\boldsymbol{J}$ *to lie parallel to* $\boldsymbol{v}$

promote the photofragment rotational excitation. These include the following possibilities:[11,13] (i) rotation of the parent molecule, (ii) the impulse caused by the O—O repulsion, (iii) vibration of the parent molecule, in the v_2'' and v_6'' symmetric and asymmetric O—H bending modes (which should produce j_{OH} perpendicular to $\boldsymbol{v}$) and v_4'', the torsional mode (which should tend to align j_{OH} parallel to $\boldsymbol{v}$), and (iv) a torsional dependence in the excited-state potential. Rotation of the parent molecule can only provide some $40 \, cm^{-1}$ of rotational excitation per OH fragment,[13] which is clearly inadequate to account for the experimental data. Nor can the impulse model completely explain the observations (though some contribution from this source cannot be ruled out) as it predicts a rotational distribution peaked sharply at $N' = 5$,[13,78] whereas the experimental distribution is fairly broad. The actual data can be explained by a combination of factors (iii) and (iv),[13] the particular weighting of bending and torsional forces being determined by a semi-classical model that varies the contributions from the two sources (subject to energetic constraints) until the bipolar moments $\beta_0^0(22)$ and $\beta_0^2(02)$ are adequately reproduced. An additional test of this model would be to fit the bipolar moments for the photodissociation of D_2O_2. Any torsional dependence on the excited-state potential should be comparable to H_2O_2, whereas the different zero-point energies and impulsive lever arm should produce a different set of vector correlations if these other motions make an important contribution to the dissociation dynamics.

A simple semi-classical model, which assumes that rotation is initially generated about an axis perpendicular to the plane containing the O—H bond in planar *trans*-H_2O_2 and which assumes the operation of Hund's case *b* coupling in the OH fragment,[13] provides qualitative agreement with the observations but fails to match the low-*j* behaviour of the bipolar moments $\beta_0^2(02)$ and $\beta_0^0(22)$ determined from the data at 266 nm (presumably as the Hund's case *b* coupling employed is not an accurate approximation). The model also over-estimates the bipolar moment $\beta_0^2(22)$, associated with the mutual correlation between ν, j_{OH}, and μ. This last point is not surprising[13] as the quantity $\beta_0^2(22)$ is the most sensitive to rotation of the parent molecule before dissociation. The recent results obtained for H_2O_2 should soon be complemented by data on nitrous acid.[7] An earlier study of HONO[75] failed to detect any probe polarization dependence of the Doppler profiles of the OH($X^2\Pi$) fragment, but recent lineshape simulations[7] using the previous experimental values of β and $\mathscr{A}_0^{(2)}$ show that the Doppler profiles should be slightly probe polarization dependent even in the absence of a (ν, j) correlation.

$(CH_3)_2NNO$ *and* $(HCO)_2$. In addition to the triatomic and tetra-atomic molecules discussed so far two larger polyatomic molecules have also been examined, namely dimethylnitrosamine[14] [$(CH_3)_2NNO$] and glyoxal[9] [*trans*-$(HCO)_2$].

The work on $(CH_3)_2NNO$ was amongst the first to recognize the effect that a (ν, j) correlation could have on Doppler-broadened lineshapes.[14] No quantitative analysis of the (ν, j) correlation was possible; indeed any analysis would have been complicated by the appreciable width of the translational-energy distribution of the NO fragments. However, it was possible to determine qualitatively that the fragment recoil and rotation were approximately perpendicular, which was interpreted as indicating a

[78] G. Ondrey, N. van Veen, and R. Bersohn, *J. Chem. Phys.*, 1983, **78**, 3732.

planar dissociation process. The last molecule to be discussed[9] provides the only example to date of a predissociated state ($\tau \sim 1\ \mu s$) that lives long enough for any correlation between the molecule and laboratory frames to be destroyed but still shows split Doppler profiles presumably caused by the (v, j) correlation created at the moment of dissociation. It is clear from Dixon's theoretical treatment[7] that bipolar moments of the form $\beta_0^0(kk)$ cannot be averaged to zero by parent molecular rotation. So in theory the $\beta_0^0(22)$ term should still contribute even if the dissociation is long delayed. Such an effect has been observed in the photodissociation of glyoxal from the 8^1 level of the first excited singlet state.[9] The data indicate a preference for v perpendicular to j_{CO}, presumably due to dissociation from a planar molecular configuration.

4   A Forward Look

The review has highlighted three major developments in the sophistication of experiments that probe the dynamics of molecular photodissociation: these include the first fully state-selected experiments, the advent of double-resonance techniques to allow state selection in systems where severe spectral-line broadening (or congestion) would otherwise prevent it, and the advent of Doppler or time-of-flight spectroscopic probing of both vector correlations and scalar distributions. The simplest forward look is to anticipate a rapid growth in experiments of each type. In particular, the possible combination of photoselection via double-resonance techniques, coupled with state-selective monitoring of photofragment distributions rather than simply the decay of the selected parent molecular state, would greatly extend the range of systems on which fully state-selected measurements could be made. The extension of time-of-flight measurements of the type initiated by Welge's group,[43] where the velocity spectrum of scattered hydrogen atoms reflects the internal-energy disposal in the molecular sister fragment, should become extremely valuable. Its special importance will be in systems where the molecular fragments are polyatomic and non-fluorescent. This last comment contains within it the otherwise unspoken thought that most molecules contain more than three or four atoms! The zone of ignorance lying between the dynamical world of small molecular photodissociations and the possibly statistical world of much larger polyatomic photodissociations was noted in an earlier review.[1] Recent experiments by Leone[79] employing infrared fluorescence spectroscopy to monitor energy disposal in the primary fragments generated from the photodissociation of propanone illustrate an alternative route to obtaining information on internal-energy disposal. The sensitivity and high resolution offered by Fourier transform infrared fluorescence spectrometers or by tunable diode laser absorption spectrometers promise rapid developments in the study of energy disposals in larger systems.

[79] D. J. Donaldson and S. R. Leone, *J. Chem. Phys.*, 1986, **85**, 817.

The success of the Doppler laser-induced fluorescence probing of CO fragments from $(HCO)_2$ by Houston and co-workers[9] in providing new information on vector correlations, despite the long predissociative lifetime, the molecular complexity, and the variety of alternative exit channels, will encourage application of this new approach more widely than might otherwise have been thought likely. The experimental community owes a considerable debt to the theoretical guidance that has been provided by Greene and Zare,[4,5] by Dixon,[7] and, before them, by Herschbach and his colleagues[76,80,81] in helping them to analyse and interpret the new vector correlation data. However, in case it may be thought that all possible correlations have now been demonstrated, an elusive one still to come is that between the rotational angular momenta of *two* molecular fragments, for example the two OH fragments generated from H_2O_2.

The results of all this experimental sophistication will be, already is, a plethora of detailed dynamical information that should challenge the theoreticians for years to come.

Acknowledgement. The Reporters are grateful to S.E.R.C. for their support of work discussed from their own laboratory, for a postdoctoral fellowship (A.H.), and for a studentship (M.P.D.). They also appreciate the help of Mrs. M. Krause in the preparation of the manuscript.

[80] D. A. Case and D. R. Herschbach, *Mol. Phys.*, 1975, **30**, 1537.
[81] J. D. Barnwell, J. G. Loeser, and D. R. Herschbach, *J. Phys. Chem.*, 1983, **87**, 2781.

CHAPTER 5

Photodissociation Processes in NO-Containing Molecules

H. REISLER, M. NOBLE, and C. WITTIG

1 Introduction

Photodissociation events in which NO is produced as a primary photofragment have recently been the subject of detailed experimental studies. In particular, the 'slow' predissociation of nitroso molecules (R—NO) and the contrasting 'fast' predissociation of nitrites (RO—NO) have been studied extensively, yielding valuable information about both dissociative and radiationless transition mechanisms. In this chapter we shall review the photo-initiated dissociation of NO-containing molecules, concentrating on the nitrosos and nitrites, but also including recent work on other similar systems (*e.g.* R_2N—NO).

Given that several of the parent molecules present considerable chemical difficulties and that quantitative detection of nascent NO quantum states is not facile, the question "Why have these systems attracted such interest?" arises. The answer lies in the fact that the parent molecules exhibit a number of interesting phenomena and that NO is in itself an interesting and informative species, with several coupled contributions to its overall angular momentum. The ground state of NO is $^2\Pi$ and conforms to Hund's case *a* at low rotational levels, becoming an intermediate *a/b* case at high *J*.[1] Consequently, two sets of information can be derived from relative spectral-line intensities. Scalar quantities such as rotational- and vibrational-energy distributions can be obtained, indicating the energy partitioning among the fragments. Vector properties such as alignment of the unpaired spin, orbital angular momentum, or overall rotation vector can be similarly obtained. Also, by using polarized lasers and sub-Doppler resolution spectroscopy, it is possible to observe anisotropy in the actual photoejection angle, measured with respect to the laboratory-fixed electric vector of the excitation laser.

---

[1] G. Herzberg, 'Molecular Spectra and Molecular Structure I. Spectra of Diatomic Molecules', Van Nostrand, New York, 1950, Ch. 5.

From the vector properties of the nascent NO, much can be deduced about both the dissociation time-scale and the nature and extent of the forces acting during dissociation. The NO $A^2\Sigma^+ \leftarrow X^2\Pi$ system gives rise to complicated LIF (laser-induced fluorescence) or MPI (multi-photon ionization) spectra when using two-photon excitation at ~ 450 nm; for one-photon LIF, excitation is achieved at a difficult wavelength (~ 225 nm). Nevertheless, the multi-faceted nature of NO makes it a most appealing diatomic fragment.

Turning to the parent molecules, nitroso compounds present 'textbook examples' of slow predissociation. The overall process consists of two distinct stages, making these molecules excellent test cases for radiationless transition and unimolecular reaction theories. The $\tilde{A}^1A'' \leftarrow \tilde{X}^1A'$ ($\pi^* \leftarrow n$) spectrum is discrete and can be rovibrationally resolved, provided the molecule is not so large as to cause congestion and overlap. After absorption of a photon, the molecule may fluoresce or undergo a radiationless transition(s) to vibrationally excited levels of the S_0 ground state ($\tilde{X}^1A'$) or the low-lying T_1 triplet state ($\tilde{a}^3A''$). Dissociation occurs on the S_0 and/or T_1 surfaces by vibrational or rovibrational predissociation mechanisms, provided the energy exceeds D_0, the dissociation energy of the weakest bond. The dissociation is slow, involving extensive energy redistribution prior to bond fission, and therefore memory of the initial excitation conditions is lost. With this type of mechanism unimolecular reaction rates can be estimated rather well using statistics, and product-state distributions can also be estimated using statistics, provided the dissociating surface has no exit channel barrier.

Conversely, the nitrites have $\tilde{A}^1A'' \leftarrow \tilde{X}^1A'$ ($\pi^* \leftarrow n$) transitions that give rise to very diffuse spectra, with barely resolved progressions in the Franck–Condon active vibrations. Their predissociation is therefore fast relative to molecular rotation ($< 10^{-12}$ s) but comparable to or slower than a vibrational period (10^{-14} s). There is not enough time for full energy randomization, and as a consequence some of the fragment vector properties show significant alignment and state distributions are very non-statistical.

When any molecular property or set of properties is studied, it is important to know how unique or general they are. One way of investigating this is to vary molecular parameters such as state densities, and several members of both the nitroso and the nitrite families have been studied, from the prototypical HNO and HONO up to $(CH_3)_3CNO$ and $(CH_3)_3CONO$. It is interesting and illuminating to compare and contrast the effects on predissociation of the variation of the 'passive' part of the molecule.

Throughout the work reviewed in this chapter, a dominant theme is the interplay between spectroscopy and dynamics. It is certainly the advent of laser spectroscopy that has made such experiments possible. For example, the details of the predissociation mechanism of HNO were completely unravelled just from a study of its laser excitation spectrum. This technique has since been applied to other interesting classes of molecules. Many of the recently studied phenomena in laser-induced photophysics and photochemistry have been observed, often for the first time, in one or more of the molecules discussed below.

2 Predissociation of Aliphatic Nitroso Compounds – RNO

Spectroscopy and Structure

The nitroso compounds share several common features in their spectroscopy, electronic structure, and photophysics. Figure 1 shows a schematic view of the three lowest potential curves of a typical nitroso compound viewed along the R—NO reaction co-ordinate. The only molecule in this class for which the potential surfaces have been accurately calculated is HNO,[2] but the basic shapes of the potentials are common to all. The ground

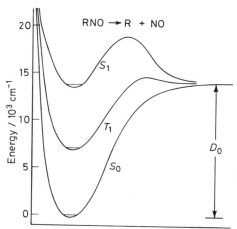

Figure 1 *A schematic diagram of the three lowest potential surfaces of a typical nitroso compound along the R—NO reaction co-ordinate*

state ($\tilde{X}^1 A'$) has no barrier to dissociation in the absence of rotation,[2] and values of D_0 range from $13\,856 \pm 79\ \text{cm}^{-1}$ for CF_3NO[3] to $17\,085 \pm 10\ \text{cm}^{-1}$ for NCNO.[4] The S_1 states ($\tilde{A}^1 A''$) have origins $11\,000$—$14\,000\ \text{cm}^{-1}$ above S_0, and S_1 correlates to ground-state products via a large barrier,[5-9] since initial correlation (to excited-state NO) undergoes an avoided crossing. Little is known about the triplet surface, and its origin and vibrational modes have been measured only for HNO.[10] Calculations for HNO place T_1 midway between S_0 and S_1, *i.e.* $\sim 7000\ \text{cm}^{-1}$ above S_0,[5,8,9]

[2] R. N. Dixon, K. B. Jones, M. Noble, and S. Carter, *Mol. Phys.*, 1981, **42**, 455.
[3] R. D. Bower, R. W. Jones, and P. L. Houston, *J. Chem. Phys.*, 1983, **79**, 2799.
[4] I. Nadler, H. Reisler, M. Noble, and C. Wittig, *Chem. Phys. Lett.*, 1984, **108**, 115.
[5] A. W. Salotto and L. Burnelle, *Chem. Phys. Lett.*, 1969, **3**, 80.
[6] P. A. Freedman, *Chem. Phys. Lett.*, 1976, **44**, 605.
[7] V. Marudarajan and G. A. Segal, *Chem. Phys. Lett.*, 1986, **128**, 1.
[8] A. A. Wu, S. D. Peyerimhoff, and R. J. Buenker, *Chem. Phys. Lett.*, 1975, **35**, 316.
[9] O. Nomura and S. Iwata, *Chem. Phys. Lett.*, 1979, **66**, 523.

in good agreement with the measured value (6280 cm^{-1}),[10] and suggest that there is a small barrier in the T_1 dissociation channel.[8,9] Qualitatively similar T_1 surfaces are anticipated for the other nitroso compounds.

The $\tilde{A}^1 A'' \leftarrow \tilde{X}^1 A'$ spectrum is discrete and can be rovibrationally resolved, provided the molecule is not so large as to cause congestion and overlap. $\tilde{A}^1 A'' \leftarrow \tilde{X}^1 A'$ spectra were partially analysed for HNO,[2,11–14] DNO,[15,16] NCNO,[17,18] CH$_3$NO,[19] CF$_3$NO,[20–22] CF$_2$ClNO,[23,24] and Bu^{t}NO.[25] In all cases but NCNO, LIF of the parent molecule was used. With NCNO, fluorescence is very weak and two-photon absorption to a dissociative state was employed.[26,27] The first photon excites the molecule via the $S_1 \leftarrow S_0$ system, and a second photon carries it to a dissociative surface yielding $CN(X^2\Sigma^+)$, which is monitored by LIF. By scanning the excitation laser while monitoring $CN(X^2\Sigma^+)$, the absorption spectrum of NCNO was obtained. In jet-cooled spectra of NCNO, the most Franck–Condon-active modes near threshold are the NCO and NCN bends (v_4 and v_5).[18] A common feature of the aliphatic R—NO compounds is their eclipsed ground state and staggered S_1 state with respect to the dihedral (torsional) angle along the C—N bond.[19,22,25] As a consequence of this change in geometry upon $\pi^* \leftarrow n$ excitation, the torsional–vibrational progression (hindered internal rotation) has a very steep Franck–Condon profile. Indeed, CF$_3$NO,[22] Bu^{t}NO,[25] and CF$_2$ClNO[23] jet-cooled spectra show a convergent progression of peaks with rapidly increasing intensity to higher energy and a maximum near the top of the torsional potential barrier. A detailed analysis has been given for CF$_3$NO.[22] Most of the dissociation studies reported to date involve excitation of members of the torsional progression.

[10] H. B. Ellis, jun. and G. B. Ellison, *J. Chem. Phys.*, 1983, **78**, 6541.

[11] F. W. Dalby, *Can. J. Phys.*, 1958, **36**, 1336.

[12] J. L. Bancroft, J. M. Hollas, and D. A. Ramsay, *Can. J. Phys.*, 1962, **40**, 322.

[13] R. N. Dixon, M. Noble, C. A. Taylor, and M. Delhoume, *Faraday Discuss. Chem. Soc.*, 1981, **71**, 125.

[14] R. N. Dixon and C. A. Rosser, *J. Mol. Spectrosc.*, 1985, **110**, 262.

[15] R. N. Dixon and C. A. Rosser, *Chem. Phys. Lett.*, 1984, **108**, 323.

[16] J. C. Petersen, *J. Mol. Spectrosc.*, 1985, **110**, 277.

[17] J. Pfab, *Chem. Phys. Lett.*, 1983, **99**, 465.

[18] M. Noble, I. Nadler, H. Reisler, and C. Wittig, *J. Chem. Phys.*, 1984, **81**, 4333.

[19] N. P. Ernsting, J. Pfab, and J. Römelt, *J. Chem. Soc., Faraday Trans. 2*, 1978, **74**, 2286.

[20] R. D. Gordon, S. C. Dass, J. R. Robins, H. F. Shurvell, and D. R. F. Whitlock, *Can. J. Chem.*, 1976, **54**, 2658.

[21] R. D. Gordon, *Int. Rev. Phys. Chem.*, 1986, **5**, 121.

[22] B. M. DeKoven, K. H. Fung, D. H. Levy, L. D. Hoffland, and K. G. Spears, *J. Chem. Phys.*, 1981, **74**, 4755.

[23] N. P. Ernsting, *J. Chem. Phys.*, 1984, **80**, 3042.

[24] J. A. Dyet, M. R. S. McCoustra, and J. Pfab, *Faraday Discuss. Chem. Soc.*, 1986, **82**, in press.

[25] M. Noble, C. X. W. Qian, H. Reisler, and C. Wittig, *J. Chem. Phys.*, 1986, **85**, 5763.

[26] H. Reisler, G. Radhakrishnan, D. Sumida, J. Pfab, J. S. Chou, I. Nadler, and C. Wittig, *Isr. J. Chem.*, 1984, **24**, 259.

[27] I. Nadler, J. Pfab, H. Reisler, and C. Wittig, *J. Chem. Phys.*, 1984, **81**, 653.

Radiationless Processes

In addition to the spectroscopic studies, nitroso compounds offer the opportunity to study non-radiative processes in a homologous series ranging from small to large molecules. Although the spectroscopy and dissociation mechanisms of nitroso compounds are rather similar, their non-radiative processes vary greatly, as expected, with the size and complexity of the molecule.

It is well known that in nitroso compounds the $\pi^* \leftarrow n$ transition is orbitally forbidden and therefore the $\tilde{A}^1A''$ radiative lifetime, $\tau_{rad.}$, is quite long ($\sim 30\,\mu s$) and similar for all nitroso molecules investigated to date.[28] The fluorescence lifetimes, however, vary greatly from molecule to molecule, and depend on the competing non-radiative processes. In HNO, levels below D_0 have fluorescence lifetimes comparable to the radiative lifetimes, despite numerous perturbations evident in the spectra.[13,29] Above D_0, those levels that are coupled to S_0 and/or T_1 dissociate very rapidly, and the fluorescence quantum yield (Φ_f) drops to zero as expected (see section on HNO).[2,13] HNO definitely belongs to the small-molecule case.

To date, fluorescence lifetimes have not been reported for NCNO, although weak, long-lived emission has been observed with 300 K samples.[27,30] Contamination by NO_2 makes lifetime measurements very difficult. However, Knee *et al.* were recently able to probe excited-state lifetimes indirectly.[31] They obtained 'two-colour spectra' in which a visible or near-i.r. ('red') photon excites the molecule via the $S_1 \leftarrow S_0$ absorption, this being followed by the absorption of a 300 nm ('blue') photon that carries the molecule to a dissociative surface that yields $CN(B^2\Sigma^+)$. The overall process is monitored via CN $B^2\Sigma^+ \rightarrow X^2\Sigma^+$ fluorescence, and by varying the delay between the 'red' and 'blue' photons the lifetime of the intermediate state can be obtained. They show that above D_0 the lifetime is very short (< 1 ns) and equal to the appearance time of CN generated via single 'red' photon absorption. Qian *et al.*[32] extended the measurements to the region below D_0 and found lifetimes $> 40\,\mu s$ just below D_0. These measurements are limited by the physical removal of excited molecules (expanded in an Ar jet) from the detection region, and thus the observed lifetimes are lower limits. Measurements carried out near the band origin ($\sim 11\,500\,\text{cm}^{-1}$) show similarly long lifetimes, without evidence of state specificity. The lack of observed state specificity may again be a result of the transport-limited nature of the experimental arrangement. For example, if the measured lifetime, $\tau_{obs.}$, obeys the relation:

$$\tau_{obs.}^{-1} = \tau_f^{-1} + \tau_{removal}^{-1} \tag{1}$$

[28] J. G. Calvert and J. N. Pitts, 'Photochemistry', Wiley, New York, 1967.

[29] K. Obi, Y. Matsumi, Y. Takeda, S. Mayama, H. Watanabe, and S. Tsuchiya, *Chem. Phys. Lett.*, 1983, **95**, 520.

[30] I. Nadler, J. Pfab, G. Radhakrishnan, H. Reisler, and C. Wittig, *J. Chem. Phys.*, 1983, **79**, 2088.

[31] L. R. Khundkar, J. L. Knee, and A. H. Zewail, *J. Chem. Phys.*, 1987, in press.

[32] C. X. W. Qian, H. Reisler, and C. Wittig, *Chem. Phys. Lett.*, in press.

where τ_f^{-1} is the fluorescence rate and $\tau_{removal}^{-1}$ is the rate of transport out of the detection region, and if $\tau_{removal} \ll \tau_f$, then differences in τ_f may not be discerned. Evidence of state specificity is obtained, however, by scanning the 'red' laser and observing the linewidths of the transitions.[18,32] Often there are striking differences in the linewidths of adjacent lines, and transitions involving excited v_4 and v_5 bending modes appear to be much broader than those with quanta of v_2 (NO stretch). Rotational structure in the broad bands could not be resolved with a reasonably narrow-bandwidth (0.04 cm^{-1}) laser. Broadening can result either from IVR in S_1 (in which case τ_f will be equal for the sharp and broad lines) or from couplings with T_1 and/or S_0, which could result in different τ_f values for the sharp and broad lines. Based on considerations of densities of states, Qian *et al.* prefer the latter interpretation.[32] Thus, in the jargon of radiationless transitions,[33] NCNO appears to belong to the small-molecule case and the molecular eigenstates, ψ_n, are given by:

$$\psi_n = a_n\varphi_{ni} + \Sigma b_{nj}\varphi_{jf} \qquad (2)$$

where φ_{ni} is the optically excited state and the φ_{jf} are states to which φ_{ni} couples. S_0 state densities are typically 50 and 280 per cm^{-1} near the S_1 origin and D_0, respectively. Thus, the number of states that can couple to a single φ_{ni} is low, and the small-molecule limit can be invoked. In this case (as with NO$_2$) there are many additional spectral lines that may emerge in the absorption spectrum, giving rise to an apparent line broadening. There are then no possibilities for the occurrence of irreversible radiationless decays. Thus, below D_0 φ_{ni} is 'diluted' by the coupled states, and consequently $\tau_f > \tau_{rad.}$. Above D_0 the molecule dissociates via the φ_{jf} components, and τ_f of the mixed state becomes much shorter than $\tau_{rad.}$. The φ_{ni} component of ψ_n probably also carries the oscillator strength for the 'blue' photon excitation, which yields CN($B^2\Sigma^+$).

In NCNO the vibrational density of states near D_0, $\rho(S_0)$, is only ~ 280 per cm^{-1}, while for CF$_3$NO and Bu^{t}NO $\rho(S_0) \sim 10^6$ and 10^{12} per cm^{-1}, respectively.[25] Thus, these molecules definitely belong to the large-molecule limit, where practically irreversible radiationless decay [via internal conversion (IC) or intersystem crossing (ISC)] is observed, rendering $\tau_f \ll \tau_{rad.}$. Indeed, in the case of jet-cooled CF$_3$NO the measured fluorescence decay times near the electronic origin are less than 200 ns, decreasing rapidly with excitation energy,[3] and Φ_f is less than 10^{-2}.[3,28] Spears and Hoffland give a detailed account of the couplings between S_1 and S_0 in CF$_3$NO and identify the R—NO torsion as a promoting mode.[34] Here, state-specific effects are apparent, and different S_1 vibrational modes exhibit different fluorescence lifetimes (Table 1). In Bu^{t}NO, as expected, the fluorescence lifetimes are even

[33] J. Jortner and R. D. Levine in 'Photoselective Chemistry', Part 1, ed. J. Jortner, R. D. Levine, and S. A. Rice, *Adv. Chem. Phys.*, 1981, **50**, 1.
[34] K. G. Spears and L. D. Hoffland, *J. Chem. Phys.*, 1981, **74**, 4765.

Table 1   *Photodissociation of jet-cooled* CF_3NO: *$\tilde{A}^1A''$-state lifetimes and NO appearance times[a]*

Vibrational mode[b]	λ/nm[b]	Vibrational energy/cm^{-1}	CF_3NO lifetime/ns	NO appearance time/ns
12^0	712.68	0.0	301 ± 66	
12^1	707.81	96.5	285 ± 38	224 ± 40
12^2	703.32	186.7	233 ± 14	
8^1	700.01	254.1		142 ± 25
12^3	699.23	269.9	193 ± 24	160 ± 30
12^4	695.51	346.5	144 ± 17	
11^1	693.26	393.1	88 ± 14	83 ± 15
7^1	692.48	409.4	122 ± 15	113 ± 20
12^5	692.16	416.1	112 ± 5	125 ± 20
6^1 or 10^1	691.39	432.1	210 ± 33	
$12^2 8^1$	690.99	440.4	106 ± 16	117 ± 20
12^{6a} and 5^1	689.89	463.5	97 ± 9	95 ± 15
12^{6e}	688.75	487.6	107 ± 2	
$12^1 7^1$	688.19	499.4	104 ± 6	
12^{7e}	687.10	522.4	107 ± 2	

[a] R. D. Bower, R. W. Jones, and P. L. Houston, *J. Chem. Phys.*, 1983, **79**, 2799. [b] B. M. DeKoven, K. H. Fung, D. H. Loy, L. D. Hoffland, and K. G. Spears, *J. Chem. Phys.*, 1981, **74**, 4755

shorter (30—40 ns near the band origin).[25] At higher energies above threshold (600—1000 cm^{-1}) Φ_f decreases abruptly, probably due to efficient mixings and Fermi resonances within S_1 that tend to promote couplings with the bath states.[25]

A particularly intriguing case that shows both large- and intermediate-molecule behaviour is CF_2ClNO.[24] Dyet *et al.* find that, while the fluorescence of levels close to the electronic origin (14 190 cm^{-1}) can be well represented by single exponential decays of about 800 ns, decays with increasingly bi-exponential character become apparent for levels above 14 400 cm^{-1}, with the lifetimes of both fast and slow components decreasing with increasing energy. The single exponential decays may derive from IC to S_0 as with the case of CF_3NO. The bi-exponential decays may represent ISC to T_1, whose vibrational-state density at this energy is only $\sim 10^3$ per cm^{-1}. The S_1–T_1 coupling here presumably corresponds to the intermediate-level structure and exhibits the typical two-exponential decay.[33] The slower component then represents delayed fluorescence due to reversible T_1–S_1 ISC.[24] ISC in CF_2ClNO may be enhanced relative to CF_3NO by the stronger spin–orbit coupling induced by the heavier Cl atom.

HNO

The prototypical nitroso molecule is HNO. This species is chemically unstable, and the unimolecular reaction HNO$^\dagger \rightarrow$ H + NO(v, J) has not been studied in terms of nascent product excitations because of experimental difficulties. However, the predissociation of electronically excited $HNO(\tilde{A}^1A'')$ can be examined by monitoring the behaviour of the parent molecule. Dixon and co-workers utilized this technique to probe the predissociation in detail and depth.[2,35] In this work, theory and experiment were successfully combined, and the LIF results were rationalized with the aid of analytical potential-energy surfaces.

Since HNO is light, the $S_1(\tilde{A}^1A'') \leftarrow S_0(\tilde{X}^1A')$ spectrum is reasonably well resolved at 300 K, and single S_1 rovibrational levels can be selectively populated without jet-cooling. This permitted effects due to parent rotation to be studied, in contrast to larger nitroso molecules. The S_0 equilibrium bond angle is 108.6°, as determined by microwave[36] and infrared spectroscopy,[37] and HNO is therefore a nearly symmetric prolate top with rotational quantum numbers J and K_a (the a-axis projection of J). The molecule belongs to the C_s point group, and all vibrations (v_1 NH stretch, v_2 NO stretch, and v_3 bend) have symmetric A' character.

Before the work of Dixon's group, the predissociation of HNO had been inferred both by line broadening in absorption $(S_1 \leftarrow S_0)$[12] and by the sudden drop of Φ_f in the chemiluminescence spectrum.[38] The latter gave an upper limit for the predissociation energy threshold as $\sim 17\,000\,\text{cm}^{-1}$. Various theoretical studies were performed in attempts to elucidate the predissociation mechanism.[5,8,9] From early studies of HNO predissociation, it had variously been concluded that predissociation involved (i) tunnelling through the $S_1(\tilde{A}^1A'')$ excited-state barrier,[5,8] (ii) mixing with levels of $T_1(\tilde{a}^3A'')$,[39] and (iii) mixing with levels of S_0.[6] Of these, mechanism (i) appeared least realistic, since all of the observed absorptions to the $\tilde{A}^1A''$ state show discrete rovibrational structure.

Dixon and co-workers studied the properties of individual $\tilde{A}^1A''$ rotational levels using the technique of high-resolution LIF of the $\tilde{A}^1A'' \leftarrow \tilde{X}^1A'$ system[2,13,14,35] and clearly established that threshold predissociation occurs via S_0. The mixing between S_1 and S_0 occurs by electron–Coriolis coupling and therefore has a rotational dependence, which they observed.[2] At reaction threshold it was assumed that the heights of the centrifugal barriers on S_0 are minimized by J being aligned with the axis of largest moment of inertia, the c-axis. By using an accurate analytical potential-energy surface for S_0 and calculating effective c-axis centrifugal barriers, the different rovibrational dissociation thresholds were predicted using $D_0 = 16\,450 \pm 10\,\text{cm}^{-1}$. In addition, they obtained data that indicated the opening of the T_1 dissociation channel at $\sim 17\,350\,\text{cm}^{-1}$ for those levels that could not predissociate via S_0 owing to symmetry restrictions.[2]

The involvement of S_0 was proved by simple but elegant measurements.[2,35] If predissociation involves $\tilde{a}^3A''$ or $\tilde{X}^1A'$, then couplings of the

[35] R. N. Dixon and M. Noble, *Springer Ser. Chem. Phys.*, 1979, **6**, 81.
[36] S. Saito and K. Takagi, *J. Mol. Spectrosc.*, 1973, **47**, 99.
[37] J. W. C. Johns and A. R. W. McKellar, *J. Chem. Phys.*, 1977, **66**, 1217.
[38] M. J. Y. Clement and D. A. Ramsay, *Can. J. Phys.*, 1961, **39**, 205.
[39] M. A. A. Clyne and B. A. Thrush, *Faraday Discuss. Chem. Soc.*, 1962, **33**, 139.

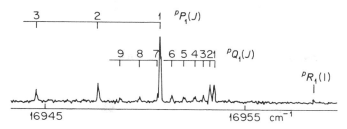

Figure 2 *LIF of the 101–000 band in the $\tilde{A}^1A''$–$\tilde{X}^1A'$ system of HNO.[2] Rotationally induced predissociation is the cause of the rapid drop in intensity with increase in J'*

(Reproduced with permission from *Mol. Phys.*, 1981, **42**, 455)

non-Born–Oppenheimer type must arise between the initially populated $\tilde{A}^1A''$ level and the $\tilde{a}^3A''$ or $\tilde{X}^1A'$ manifolds to allow a radiationless transition. The singlet states have different electronic symmetry (A' and A'') and can only be coupled by nuclear motions of A'' symmetry. The only such motions are *a*- and *b*-axis rotation. Conversely, S_1 and T_1 both have A'' electronic symmetry and can be mixed by spin–orbit coupling with no specific rotational dependence. Therefore, if predissociation occurs via S_0, as opposed to T_1, the $J = 0$ levels of $\tilde{A}^1A''$ should be stable to predissociation, even if their energy exceeds the dissociation threshold. LIF signals were detected for two $\tilde{A}^1A''$ vibronic states above D_0 and were found to consist of only one strong line in each case [the $^PP_1(1)$ line terminating in $J' = K_a' = 0$] as shown in Figure 2, which displays the LIF spectrum of the 101 ← 000 band. This demonstrates that rotation is the prerequisite for predissociation, proving that for $E^\dagger \leqslant 900\ cm^{-1}$ dissociation proceeds via S_0 rather than T_1. The extent of coupling between $\tilde{X}^1A'$ and $\tilde{A}^1A''$ was further evidenced by the large number of excited-state perturbations observed in the spectrum, which showed no magnetic susceptibility and which were assigned as $\tilde{A}$–$\tilde{X}$ mixing.[13,40] Figure 3 shows a typical perturbed rotational sub-band of the HNO LIF spectrum, in which many lines are split with anomalous intensities, even below the predissociation onset, which occurs at $J' = 12$ for this 100–000 ($K_a' = 4$) ← ($K_a'' = 3$) sub-band. Similar results were obtained in the LIF study of DNO.[16,41]

The $\tilde{A}^1A''$ and $\tilde{X}^1A'$ states of HNO form a Renner–Teller pair that arises from a degenerate $^1\Delta$ configuration of the linear molecule. The largest non-Born–Oppenheimer term coupling $\tilde{A}$ and $\tilde{X}$ must be Renner–Teller coupling, *i.e.* *a*-axis electron–rotational coupling. Since the two states are degenerate in a linear configuration, mixing will be enhanced by excitation of the bending vibration.[13] Dixon and co-workers obtained evidence in support of this in the case of HNO, and the role of the bending vibration as a promoting mode for internal conversion is a common characteristic of the

[40] R. N. Dixon and M. Noble, *Chem. Phys.*, 1980, **50**, 331.
[41] R. N. Dixon and C. A. Rosser, *Philos. Trans. R. Soc. London, Ser. A*, 1982, **307**, 603.

photophysics of nitroso molecules. Dixon and co-workers were also able to demonstrate that, following coupling between S_1 and S_0, rapid rovibrational energy redistribution could occur. The internal energy of a molecule on a particular electronic surface can be divided into $E_{vib.}$ and $E_{rot.}$, and conservation of angular momentum ensures that $E_{rot.}$ is reduced as the bond lengthens, thus changing the effective potential. This is most noticeable in diatomic hydrides, as discussed by Herzberg,[1] but had not previously been observed for a polyatomic molecule. These so-called effective centrifugal barriers on the potential-energy surface are therefore dependent on the composition of E_{total} in terms of $E_{vib.}$ and $E_{rot.}$. However, HNO predissociation does not have a marked dependence on $E_{rot.}$.

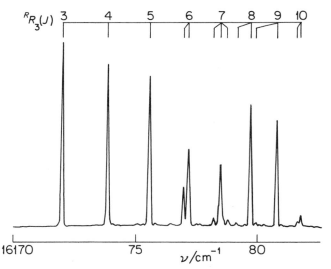

Figure 3 $^RR_3$ 100–000 *excitation spectrum of* HNO *(0.1 cm⁻¹ resolution)*.[13] (Reproduced with permission from *Faraday Discuss. Chem. Soc.*, 1981, **71**, 125)

The calculated surfaces of Dixon *et al.*[2] show that prior to dissociation there may be large-amplitude NO stretching vibrations on $\tilde{X}^1A'$, which is attractive to long range (~ 5 Å). The amplitude of these vibrations is such that the principal axes of the molecule are significantly tilted during the vibrational period and vibration and rotation are coupled by a large Coriolis effect. As a result, K_a is not a good quantum number and J can align about any molecular axis. Since $I_c > I_b \gg I_a$, the minimum value of $E_{rot.}$ will be given when J is aligned about the c-axis, and this can greatly reduce the height of the threshold centrifugal dissociation barrier. In other words:

$$E_{total} \simeq E_{vib.} + AK_a^2 + BJ(J + 1) \tag{3}$$

and, as K_a goes to zero by Coriolis-induced realignment of J, the rotational-energy term AK_a^2 disappears with a concomitant increase in $E_{vib.}$, *i.e.* at threshold AK_a^2 can be considered part of $E_{vib.}$ rather than $E_{rot.}$. This is important, since AK_a^2 is the largest rotational-energy contribution. Using the assumption of extensive Coriolis mixing, Dixon and co-workers calculated c-axis centrifugal barriers on their $\tilde{X}^1 A'$ surface for all rovibrational states whose predissociation onsets had been observed (by break-offs in the LIF spectra as Φ_f goes to zero; see Figure 2). By choosing $D_0 = 16\,450 \pm 10$ cm^{-1}, the model was shown to be in exact agreement with all their data (Figure 4) and could also explain all previous HNO predissociation data.[2] The need to be aware of the composition and fate of both $E_{vib.}$ and $E_{rot.}$, together with the conservation of J, is a theme which recurs in the case of NCNO predissociation.

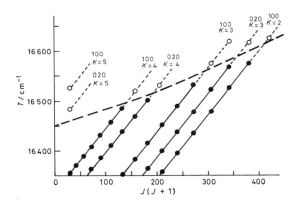

Figure 4 *The highest levels observed in LIF ($\bullet$) and the lowest missing levels ($\circ$) in the 020–000 and 100–000 vibronic transitions of the HNO $\tilde{A}^1 A''$–$\tilde{X}^1 A'$ band system.[2] The dashed curve (– – –) indicates the J-dependent threshold for predissociation calculated using the model described in the text. A dissociation energy of 16 450 cm^{-1} has been assumed in fitting this curve to the experimental data*
(Reproduced with permission from *Mol. Phys.*, 1981, **42**, 455)

The full picture of HNO threshold predissociation of an initially populated $\tilde{A}^1 A''$ level involves (i) electronic–Coriolis couplings into the $\tilde{X}^1 A'$ manifold, (ii) extensive energy flow via Coriolis couplings until $E_{vib.}$ is all localized in the NH stretch and J is aligned along the c-axis, and (iii) N—H bond rupture. Obviously, beyond the threshold region the requirements that all of $E_{vib.}$ is localized in the NH stretch and J is aligned along the c-axis are no longer stringent. The threshold mechanism has been further confirmed by recent work on DNO predissociation, and to reconcile theory and experiment it was only necessary to take account of the increased mass.[16,41]

In addition to unravelling the threshold predissociation mechanism, Dixon's group showed that for $E^\dagger \geq 900$ cm^{-1} even $\tilde{A}^1 A''$ levels with $J = 0$ can predissociate. They suggested that this is probably due to predissociation via T_1, whose dissociation barrier would therefore be ~ 900 cm^{-1}.[2]

NCNO

In the predissociation of nitroso compounds it is assumed that extensive IVR takes place on S_0 or T_1 until sufficient energy is concentrated in the weak R—NO bond to cause dissociation. Vibrational energy is presumed to flow randomly through the molecule, erasing any memory of preferential excitation caused by the coupling with S_1. If the molecule then dissociates via a loose transition state, both the reaction rates and product quantum-state distributions should reflect complete energy randomization in the dissociating state. There are several statistically based theories that predict the outcomes of such reactions. RRKM[42-44] and the statistical adiabatic channel model (SACM)[45-50] are well tested theories used extensively to predict unimolecular reaction rates. The main theories available to test product-state distributions are phase space theory (PST)[51-55] and SACM,[45-50] which calculate the outcomes based on statistical weights while conserving E and total angular momentum. Prior distributions,[56] which conserve only energy, are often applicable to larger molecules where angular momentum constraints are not important.

In order to examine the ability of statistical theories to determine product-state distributions, Wittig and co-workers chose to study NCNO (nitrosyl cyanide). This molecule behaves in many ways like other nitroso molecules, having an $\tilde{A}^1A'' \leftarrow \tilde{X}^1A'$ spectrum in the near-i.r. and visible regions that is at least partly predissociative.[17,18] However, it has the distinct advantage that both fragments, CN and NO, can be studied quantitatively by the reliable, almost routine, technique of LIF using wavelengths not significantly absorbed by NCNO. As with all nitroso species, the geometry changes accompanying $\pi^* \leftarrow n$ electronic excitation ensure that transitions to a large energy region of the upper state are Franck–Condon allowed. As a result it proved possible to study product E,V,R,T excitations in fine detail with $E^\dagger$ in the range 0—5000 cm^{-1}.[4,27,57-61] Even in such a small molecule, because of low-frequency bending vibrations it was necessary to use jet-cooled samples to prevent 'hot-band' effects that could dominate near the

[42] P. J. Robinson and K. A. Holbrook, 'Unimolecular Reactions', Wiley, London, 1972.

[43] W. Forst, 'Theory of Unimolecular Reactions', Academic Press, New York, 1973.

[44] I. W. M. Smith, 'Physical Chemistry of Fast Reactions', Vol. 2, ed. I. W. M. Smith, Plenum Press, New York, 1980.

[45] M. Quack and J. Troe, *Ber. Bunsenges. Phys. Chem.*, 1974, **78**, 240.

[46] M. Quack and J. Troe, *Ber. Bunsenges. Phys. Chem.*, 1975, **79**, 170.

[47] M. Quack and J. Troe, *Int. Rev. Phys. Chem.*, 1981, **1**, 97.

[48] M. Quack and J. Troe, *Ber. Bunsenges. Phys. Chem.*, 1977, **81**, 329.

[49] J. Troe, *J. Chem. Phys.*, 1981, **75**, 226.

[50] J. Troe, *J. Chem. Phys.*, 1983, **79**, 6017.

[51] P. Pechukas and J. C. Light, *J. Chem. Phys.*, 1965, **42**, 3281.

[52] P. Pechukas, C. Rankin, and J. C. Light, *J. Chem. Phys.*, 1966, **44**, 794.

[53] J. C. Light, *Faraday Discuss. Chem. Soc.*, 1967, **44**, 14.

[54] C. E. Klotz, *J. Phys. Chem.*, 1971, **75**, 1526.

[55] C. E. Klotz, *Z. Naturforsch., Teil A*, 1972, **27**, 553.

[56] R. D. Levine and J. L. Kinsey in 'Atom–Molecule Collision Theory – A Guide for the Experimentalist', ed. R. B. Bernstein, Plenum Press, New York, 1979, Ch. 22.

[57] I. Nadler, M. Noble, H. Reisler, and C. Wittig, *J. Chem. Phys.*, 1985, **82**, 2608.

reaction threshold.[27] Their monoenergetic data represent the most extensive and detailed set of product distributions following a unimolecular reaction reported to date.[60,61]

The results of this study can be summarized as follows. NCNO predissociates following $S_1 \leftarrow S_0$ excitation, provided that $h\nu$ exceeds D_0 ($17\,085 \pm 5\,cm^{-1}$).[4] Spectroscopic perturbations strongly indicate that vibrational predissociation near D_0 occurs on S_0 and/or T_1.[18] Calculations indicate that the barrier to direct $\tilde{A}$-state dissociation may be at least as high as $5000\,cm^{-1}$.[7] The terminal bending vibrations, ν_4 and ν_5, are found to be efficient promoting modes for S_1–S_0 mixing, as expected for states that are a Renner–Teller pair ($\tilde{A}^1A''/\tilde{X}^1A'$) and are therefore degenerate for a linear molecule.[18] NCNO has an equilibrium configuration that is *trans* and planar in both the S_1 and S_0 states.[18] At the dissociation threshold there is no detectable product translational or rotational excitation, confirming the absence of a barrier to dissociation. Products appear on the sub-ns timescale except at threshold, where finite rates (20—40 ns) are suggestive of a small centrifugal barrier and possible tunnelling.[4] The barrier height and indirect estimates of the rotational constant for the transition state indicate that it is very loose.

When product vibrations are energetically inaccessible, the rotational distributions of both CN and NO are fitted quite accurately using PST.[57,58,61] Figure 5 shows the agreement between theory and experiment for $E^\dagger = 0$—$939\,cm^{-1}$. The photodissociation spectrum for NCNO in this region is also shown, with an abrupt onset at 585.3 nm ($17\,085\,cm^{-1}$). The spectrum is very dense at these wavelengths, with presumably extensive couplings to S_0 and/or T_1, and possibly Fermi resonances within S_1, since the S_1 origin is at $11\,339\,cm^{-1}$.

As $E^\dagger$ increases, product vibrations become accessible (the $v'' = 1$ energies of NO and CN are 1876 and $2043\,cm^{-1}$, respectively), and for $E^\dagger > 1900\,cm^{-1}$ deviations from PST can be seen, *i.e.* PST overestimates rotational excitation and underestimates vibrational excitation compared to the experimental observations.[57,61] Wittig and co-workers[58] examined PST and included a restriction on energy flow between parent vibration and rotation, thus making the model more realistic. This modified form of PST was termed the separate statistical ensembles (SSE) method and is discussed in more detail below. Up to the highest $E^\dagger$ possible in these experiments ($5000\,cm^{-1}$) all product V,R distributions fit the SSE predictions to within the experimental error.[57,61] The CN spin–rotation states are equally populated, indicating no preference of spin alignment in the molecular frame. The Λ-doublet states of NO (coupling of orbital and rotational angular

[58] C. Wittig, I. Nadler, H. Reisler, M. Noble, M. Catanzarite, and G. Radhakrishnan, *J. Chem. Phys.*, 1985, **83**, 5581.

[59] J. Pfab, J. Häger, and J. Krieger, *J. Chem. Phys.*, 1983, **78**, 266.

[60] J. L. Knee, L. R. Khundkar, and A. H. Zewail, *J. Phys. Chem.*, 1985, **89**, 4659.

[61] C. X. W. Qian, M. Noble, I. Nadler, H. Reisler, and C. Wittig, *J. Chem. Phys.*, 1985, **83**, 5573.

momenta) are equally populated, which is not surprising since NO corresponds to either Hund's case *a* or *a/b* at the *J* values observed.[61] However, the NO spin–orbit states are consistently 'colder' than predicted from PST by a factor of 0.6.[61] A similar anomaly was observed by Houston and co-workers[3] when comparing the NO from CF_3NO with simple statistical (Prior) calculations. It was suggested by Wittig's group[4] that the 'hidden' weighting factor arises from angular momentum coupling as the transition from the parent wavefunction basis-set to the product wavefunction basis-set occurs during dissociation.

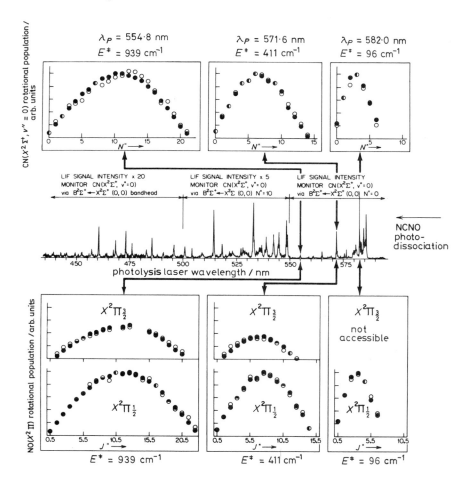

Figure 5 *A composite overview of the* NO *and* CN *rotational distributions from* NCNO *photolysis* (○ *experiment,* ● *theory*).[61] *The central trace shows the photodissociation spectrum of jet-cooled* NCNO. *The* NO *and* CN *rotational populations result from dissociation at several of the absorption maxima. The* NO *spin–orbit ratio is fixed at the experimental value*

(Reproduced with permission from *J. Chem. Phys.*, 1985, **83**, 5573)

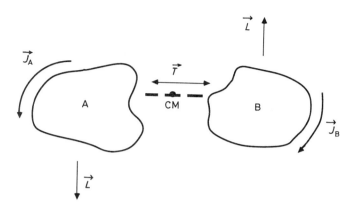

Figure 6 *Schematic representation of the loose transition state associated with* PST.[58]
Species A *and* B *contain vibrational excitation*
(Reproduced with permission from *J. Chem. Phys.*, 1985, **83**, 5581)

The SSE method has been described in detail,[57,58] but since it may have quite general applicability important aspects and concepts will be reviewed here. One difficulty with PST is that it contains implicit assumptions that allow energy to flow freely between parent vibration and rotation. In photodissociation processes, particularly using jet-cooled samples, the parent excitation may not be equally divided among vibrational and rotational degrees of freedom. In the specific case of jet-cooled NCNO following IC and/or ISC, the parent molecule is vibrationally excited but rotationally quite cold. PST treats the loosest of all transition states, and implicit in the model is energy randomization amongst *all* available degrees of freedom. As shown in Figure 6, the total angular momentum J_0 is the resultant of J_A and J_B (the angular momenta of the two fragments) and L, the angular momentum of the orbital motion of A relative to B. PST equilibrates all degrees of freedom while conserving $E^\dagger$ and J_0. Large L is offset by J_A and J_B, thus conserving small J_0. The SSE method contains a modification that deals explicitly with cases where energy does not flow freely between parent vibrations and rotations. To deal with this, two ensembles are considered in turn. Firstly, product vibrations are estimated from an ensemble of parent degrees of freedom, which excludes parent rotation. Secondly, product R, T excitations are estimated for each set of product vibrational states from an ensemble of parent degrees of freedom, which includes those parent vibrations that are disappearing as well as parent rotations. At this point several statistical methods can be used (PST, SACM, Prior, *etc.*), as the important physics has already been built in. For the case of NCNO, $6V$ and $3R$ parent degrees of freedom are converted to $2V$, $4R$, and $3T$ product degrees of freedom, and differences between PST and SSE are apparent when $E^\dagger$ is greater than the lowest product vibrational energy.

Good agreement between SSE and experiment is obtained for all of the observed CN and NO vibrational and rotational distributions up to $E^\dagger =$

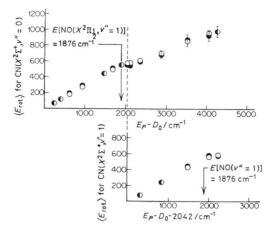

Figure 7 *Experimental* (●) *and calculated (SSE)* (○) *values of* $<E_{rot.}>$ *for* CN($X^2\Sigma^+$) *obtained by NCNO photolysis.*[58] $E_p - D_0(v'')$ *is the excitation energy minus the threshold for formation of* CN($X^2\Sigma^+$) *in the* v'' *level. The upper and lower entries are for* $v'' = 0$ *and 1, respectively. Vertical arrows at* 1876 cm^{-1} *mark the energy threshold for the* NO($X^2\Pi_{\frac{1}{2}}$, $v'' = 1$) *channel* (Reproduced with permission from *J. Chem. Phys.*, 1985, **83**, 5581)

5000 cm^{-1}.[57,61] This is illustrated in Figure 7, which shows the calculated and experimental values of the average rotational energy, $<E_{rot.}>$, of the CN fragment at various excitation energies. The fit with the vibrational distributions is just as good, as can be seen in Table 2.

Table 2 *Relative product vibrational populations for different energies in excess of the NCNO reaction threshold*[a]

$E^{\dagger}$/cm^{-1}	Experiment	PST[b]	SSE[b]
	[CN($X^2\Sigma$,$v'' = 1$)]/$\sum_{v''}$ [CN($X^2\Sigma$,v'')]		
2348	0.07 ± 0.02	0.034	0.08
2875	0.16 ± 0.02	0.11	0.17
3514	0.20 ± 0.03	0.17	0.22
4050	0.24 ± 0.03	0.20	0.24
4269	0.27 ± 0.04	0.21	0.26
	[NO($X^2\Pi$,$v'' = 1$)]/$\sum_{v''}$ [NO($X^2\Pi$,v'')]		
2348	0.12 ± 0.03	0.07	0.14

[a] I. Nadler, M. Noble, H. Reisler, and C. Wittig, *J. Chem. Phys.*, 1985, **82**, 2608; C. X. W. Qian, M. Noble, I. Nadler, H. Reisler, and C. Wittig, *J. Chem. Phys.*, 1985, **83**, 5573. [b] Values shown are for $J_0 = 5$; for $J_0 = 0$ or 10 there is only a very small change

Recently, Qian *et al.*[62] carried out experiments that probed correlations between the internal energies of the two fragments [*e.g.* to find the corresponding NO states for a specific $CN(X^2\Sigma^+,v'',J')$]. From energy and angular momentum conservation:

$$E_{\text{trans.}} = E^\dagger - [E_{\text{vib.}}(\text{NO}) + E_{\text{rot.}}(\text{NO}) + E_{\text{elect.}}(\text{NO})$$
$$+ E_{\text{vib.}}(\text{CN}) + E_{\text{rot.}}(\text{CN})] \tag{4}$$

$$J_0 = J_{\text{CN}} + J_{\text{NO}} + L \tag{5}$$

For each state of CN there is a distribution of NO states and consequently a distribution of translational energies. In the statistical theories these distributions are subject only to the constraints of the models. However, if there are dynamical biases associated with the correlations between the quantum-state distributions of the fragments, this can cause deviations from statistical predictions. For example, the distribution of NO states associated with a *specific* CN state may be non-statistical, even though the separate CN and NO state distributions are quite statistical. In order to check for such correlation, Qian *et al.* recorded LIF spectra of CN fragments with sub-Doppler resolution. Since the recoil velocity distribution is expected to be isotropic, it is possible to compare the experimental lineshape with the calculated profiles obtained using a statistical distribution of NO internal states. Typical lineshapes are shown in Figure 8 as a function of $E^\dagger$ for

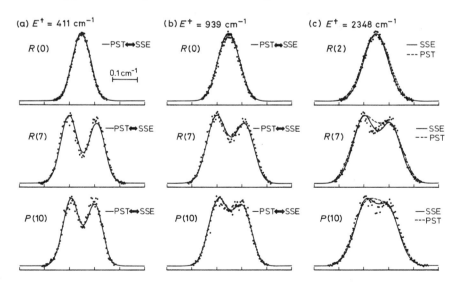

Figure 8 *Doppler profiles of $CN(X^2\Sigma^+)$ following NCNO photolysis at energy excesses:* (a) $411\,\text{cm}^{-1}$, (b) $939\,\text{cm}^{-1}$, (c) $2348\,\text{cm}^{-1}$. *The two partially resolved peaks are the spin–rotation branches (F_1 and F_2) of $CN(X^2\Sigma^+)$*[62]

[62] C. X. W. Qian, M. Noble, A. Ogai, H. Reisler, and C. Wittig, to be published.

several R and P transitions. $R(7)$ was chosen because the two spin–rotation branches (F_1 and F_2) were not completely resolved, and therefore the valley between the two peaks could serve as a sensitive measure of the fit (see Figure 8). Notice in particular the results for $E^\dagger = 2348\,\text{cm}^{-1}$, where $NO(v'' = 1)$ is already populated, and thus PST and SSE predict slightly different profiles. Clearly, the results are a better fit with SSE than PST, although most people would consider this splitting hairs. Thus, the $CN(X^2\Sigma^+)$ and $NO(X^2\Pi)$ rotational distributions are only correlated as per the statistical predictions.

CF_3NO *and* CF_2ClNO

CF_3NO was the first nitroso compound for which the NO fragment was directly monitored.[3,63-65] The NO appearance time is equal to the CF_3NO fluorescence lifetime in both 300 K and jet-cooled samples (Table 1).[3,64] Near threshold the appearance times are much longer than those calculated using RRKM theory for reaction on S_0. Thus, the lifetimes reflect the radiationless transition rates rather than the unimolecular reaction rates. Spears and Hoffland used radiationless transition rate theory to calculate the relative changes in the 300 K radiationless rates as a function of the excitation mode and energy near the S_1 origin.[34] They found better agreement between theory and experiment if IC to S_0 was involved rather than ISC to T_1. They suggested therefore that the predissociative state is S_0, at least near threshold.

Energy distributions in NO following 690—714 nm excitation have been studied in detail using jet-cooled samples.[3,65] Previous work with 300 K samples at 600—680 nm yielded similar results, except for the vibrational distributions, which were contaminated by contributions from two-photon absorptions.[3,64] The NO rotational-energy distributions obtained with excitation energies near D_0 (13 856 $\pm$ 79 cm^{-1}) are consistent with a statistical distribution of energy (Prior) in the products.[3,64] Thus, although the dissociation rate is much faster than the radiationless transition rate, all memory of the initial S_1 excitation is erased, and energy randomization on the dissociative surface appears to be complete before dissociation. As more highly excited vibronic states of S_1 are selected, the product-energy distributions begin to show signs of non-statistical behaviour, although the deviations are small.[3] These deviations may reflect a greater involvement of T_1 (see the sections on HNO and Bu^{t}NO). In contrast, the fraction of NO formed in the upper $^2\Pi_{\frac{3}{2}}$ state relative to the $^2\Pi_{\frac{1}{2}}$ state is considerably smaller than the statistical prediction, even near threshold, as with the case of NCNO.[3]

Work similar to that carried out with CF_3NO has recently been reported for CF_2ClNO (chlorodifluoronitrosomethane).[24] This molecule absorbs at

[63] M. Asscher, Y. Haas, M. P. Roellig, and P. L. Houston, *J. Chem. Phys.*, 1980, **72**, 768.
[64] M. P. Roellig, P. L. Houston, M. Asscher, and Y. Haas, *J. Chem. Phys.*, 1980, **73**, 5081.
[65] R. W. Jones, R. D. Bower, and P. L. Houston, *J. Chem. Phys.*, 1982, **76**, 3339.

485—705 nm, and jet-cooled fluorescence excitation spectra of the $\tilde{A}^1 A'' \leftarrow \tilde{X}^1 A'$ system were recorded at 670—710 nm. Preliminary results by Dyet *et al.*[24] indicate that non-radiative decay rates are relatively slow near threshold and show evidence of both IC and ISC. The NO yield spectrum obtained at 660—710 nm is very similar to the LIF spectrum, indicating no state-specific effects in the dissociation. The NO rotational-energy distributions obtained at a number of narrow features in the fluorescence excitation spectrum appear to be statistical. However, the $[^2\Pi_{\frac{3}{2}}]/[^2\Pi_{\frac{1}{2}}]$ population ratio is smaller than statistical, similar to observations in other nitroso compounds.[24]

Bu^{t}NO

As indicated above, the work on small nitroso compounds could not distinguish experimentally between dissociation on S_0 and T_1, and photodissociation experiments with 2-methyl-2-nitrosopropane [$(CH_3)_3CNO$, hereafter referred to as Bu^{t}NO] were carried out mainly to clarify this point.[25,66,67] Bu^{t}NO is structurally similar to CF_3NO but has many more vibrational degrees of freedom (39 as against 12). Consequently, the S_0 and T_1 surfaces of Bu^{t}NO have much higher densities of states, and this results in faster radiationless decay of S_1 levels than for the case of CF_3NO.[25] On the other hand, RRKM theory predicts that unimolecular reaction is considerably slower (*e.g.* several μs for dissociation on S_0 at 500 cm^{-1} above D_0),[25,66] and this permits temporal separation of the radiationless decay and the S_0 unimolecular reaction rates. Recently, the predissociation of jet-cooled Bu^{t}NO following laser excitation in the $\tilde{A}^1 A'' \leftarrow \tilde{X}^1 A'$ system has been studied in both the frequency and time domains, and it has been possible to measure separately radiationless transition rates and unimolecular reaction rates in real time.[25,67] Five types of experiments were carried out: (i) LIF spectra of the Bu^{t}NO parent, (ii) fluorescence lifetimes of Bu^{t}NO, (iii) NO yield spectra (*i.e.* photodissociation spectra), (iv) NO appearance times, and (v) LIF spectra of nascent NO.

As described above, the Bu^{t}NO fluorescence lifetimes (and hence the radiationless transition times) are short, less than 45 ns.[25] The NO appearance times, however, show two distinct regimes. At $E^\dagger < 650$ cm^{-1} ($D_0 = 13\,938 \pm 30$ cm^{-1}, S_1 origin $= 13\,911$ cm^{-1}) lifetimes greater than 3.5 μs are observed, whereas at higher $E^\dagger$ a fast risetime component (less than 15 ns, the time resolution of the experiment) becomes apparent and soon dominates.[25] Over a narrow $E^\dagger$ region near the threshold for the appearance of the fast component, a double-exponential appearance curve is observed, but the relative amplitude of the slow component decreases rapidly with excess energy. The 'fast' and 'slow' NO contributions can be seen clearly in Figure 9. Dissociation spectra were recorded by using a pump/probe configuration, fixing the probe laser wavelength at the P_{11} bandhead of NO

[66] H. Reisler, F. B. T. Pessine, Y. Haas, and C. Wittig, *J. Chem. Phys.*, 1983, **78**, 3785.
[67] M. Noble, C. X. W. Qian, H. Reisler, and C. Wittig, *J. Chem. Phys.*, 1986, **84**, 3573.

(226.34 nm, $J'' \sim 8.5$) and detecting NO via one-photon LIF. Figure 9 shows spectra obtained with pump/probe delays of ~ 60 ns and 3.2 μs, compared to the LIF spectrum of Bu^{t}NO in the same excitation region.[25,67] The spectra show clearly a group of peaks at wavelengths greater than 685 nm that have a long appearance time.

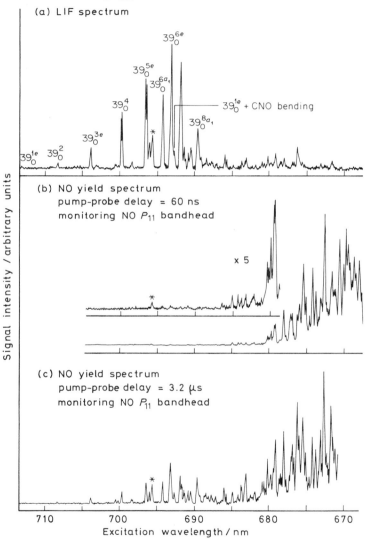

Figure 9 Bu^{t}NO *LIF and* NO *yield spectra as a function of the photolysis laser wavelength* (3/500 Torr *in* He).[25] *(a) is* Bu^{t}NO *LIF with spectroscopic assignments indicated where known; (b) and (c) are the* NO *photodissociation yield spectra recorded with the probe laser set to the* P_{11} *bandhead of* NO *at 226.3 nm, with laser delays of 60 ns and 3.2 μs, respectively. All spectra are corrected for laser energy*

(Reproduced with permission from *J. Chem. Phys.*, 1986, **85**, 5763)

Although the 'fast' and 'slow' appearance rates could easily be observed, accurate dissociation rates could not be obtained. Measurements of the slow rates were limited by removal of the molecules in the supersonic jet from the region of the probe laser. Thus, only lower limits on the dissociation lifetimes were derived. The measurements of the fast rates at $E^\dagger > 650\ cm^{-1}$ were limited by the time resolution of the experiment (~ 15 ns).[25]

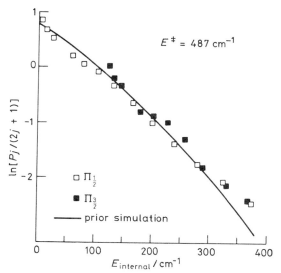

Figure 10 *Nascent NO rotational-state distributions ($\Pi_{\frac{1}{2}}$ and $\Pi_{\frac{3}{2}}$) following Bu^{t}NO photolysis at 693.67 nm (39^{6e}_0 transition) compared to a Prior calculation using $E^\dagger = 487\ cm^{-1}$. Note that the curves for the two spin states of NO coincide[25]*

(Reproduced with permission from *J. Chem. Phys.*, 1986, **85**, 5763)

The experimental results suggest that two separate processes are responsible for NO formation.[25] At $E^\dagger < 650\ cm^{-1}$ dissociation proceeds only on S_0. However, at $E^\dagger > 650\ cm^{-1}$ the barrier to dissociation via T_1 is exceeded and direct dissociation via this channel competes with dissociation on S_0. This mechanism is further supported by NO($X^2\Pi$) rotational-energy distributions. At $E^\dagger < 650\ cm^{-1}$ dissociation on S_0 leads to nascent NO distributions that can be fitted using statistical partitioning of $E^\dagger$ with a loose transition state and no exit channel barrier (Figure 10).[25] The spin–orbit population ratio, $[^2\Pi_{\frac{3}{2}}]/[^2\Pi_{\frac{1}{2}}]$, is statistical as well. Conversely, above the small ($\sim 650\ cm^{-1}$) barrier to dissociation on T_1, non-statistical behaviour is observed (Figure 11).[11] The NO rotational distributions are colder than statistical, and, most notably, the spin–orbit population ratio is much smaller than statistical. The results indicate that the two processes $T_1 \rightarrow S_0$ and $T_1 \rightarrow$ products occur at comparable rates near the top of the T_1 barrier. This causes the bi-exponential NO appearance rates, since some molecules

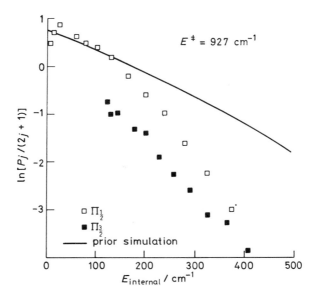

Figure 11 *Nascent* NO *rotational-state distributions* ($\Pi_{\frac{1}{2}}$ *and* $\Pi_{\frac{3}{2}}$) *following* Bu^{t}NO *photolysis at* 673.09 nm *compared to a Prior calculation using* $E^\dagger = 927$ cm^{-1}. *Notice that the* NO *spin states are anomalously populated,* i.e. $[\Pi_{\frac{3}{2}}] \ll [\Pi_{\frac{1}{2}}]^{25}$

(Reproduced with permission from *J. Chem. Phys.*, 1986, **85**, 5763)

dissociate on T_1 very quickly (less than 20 ns) whereas others remain trapped on S_0 for up to several μs. Variations in the relative amplitudes of the fast and slow components probably reflect variations in the $S_1 \rightarrow T_1$ and/or $T_1 \rightarrow S_0$ rates. At the top of the T_1 barrier, NO is internally cold due to the smaller 'effective' $E^\dagger$ above the barrier, compared with dissociation on S_0. However, *because* of the barrier, the products gain kinetic energy in the exit channel. The recoil energy will appear principally in the form of translation (plus some rotation) and will be partitioned between the fragments according to conservation of momentum, not densities of states. Thus, Bu^{t}NO offers a unique possibility for studying a simple bond fission reaction above a barrier.

The colder than statistical NO rotational distributions observed at photolysis wavelengths less than 685 nm suggest that the energy available for partitioning among internal degrees of freedom may be lower than that calculated by $E^\dagger = h\nu - D_0$. In order to check whether the energy released upon crossing the T_1 barrier might be unavailable for rotational excitation, the rotational and spin–orbit distributions that correspond to $E^\dagger = h\nu - D_0 - E_b$, where E_b is the barrier height for dissociation on T_1, were calculated. There is better agreement with the experimental distributions when the rotational distributions are calculated in this manner, suggesting that a large fraction of the potential energy released beyond the barrier is indeed channelled into translation. However, the spin–orbit population ratio still does not match the experimentally observed one.

Well above the T_1 barrier most of the Bu^{t}NO molecules that have undergone ISC into T_1 will dissociate on T_1 instead of undergoing ISC to S_0, since $T_1 \rightarrow$ products becomes faster than $T_1 \rightarrow S_0$. RRKM calculations yield T_1 dissociation times less than 10 ns when the T_1 origin is greater than 5000 cm^{-1} above the S_0 origin.[25,66] This is in accord with results on HNO, which suggest that the T_1 origin is halfway between S_0 and S_1 and that it has a small barrier to dissociation.[5,8,9]

In conclusion, Bu^{t}NO is the largest aliphatic nitroso compound whose predissociation dynamics have been studied in detail. As such, it presents not merely another test case for statistical theories but also a case for which the radiationless transitions and unimolecular reaction rates can be separated in time. Indeed, it has proved possible to observe ISC and/or IC and competitive dissociation on two surfaces (S_0 and T_1), the latter having a small barrier to dissociation. Unfortunately, because of the small excess-energy range for which the unimolecular reaction rates could be measured, meaningful comparisons with RRKM calculations could not be made.

Summary

The two aspects of the photodissociation of aliphatic nitroso molecules that provide a rather detailed and complete picture are the energy distributions in the photofragments and the coupling between the initially excited S_1 state and the S_0 and/or T_1 dissociating state(s). The latter shows great variations among the nitroso molecules, which range from the small- to large-molecule cases, depending on the state densities. Thus, the molecular eigenstates can be described using either perturbation theory (HNO, NCNO) or radiationless transition theory for the case of essentially irreversible decay (CF$_3$NO, Bu^{t}NO). The molecular eigenstates of the dissociative states of HNO and NCNO presumably still have some S_1 character. Nevertheless, it appears that the nature of the coupling between S_1 and S_0/T_1, as well as the extent of S_1 character in the dissociating state, does not bias energy partitioning in the products, as long as there are no barriers in the dissociating surface. In fact, the product-energy distributions derived from NCNO photolysis can serve as a test case for statistical theories. The complete set of product quantum-state distributions obtained for both NO($X^2\Pi$) and CN($X^2\Sigma^+$) provided the experimental data base required for development of the SSE modification to PST, as described above. The only non-statistical behaviour is the $[^2\Pi_{\frac{3}{2}}]/[^2\Pi_{\frac{1}{2}}]$ population ratio in NO($X^2\Pi$), which is often colder than statistical. When a barrier to dissociation does exist (e.g. on T_1 in the case of Bu^{t}NO), non-statistical product-energy distributions are observed.

Further work is still needed in two areas: the nature of the dissociating state and the rates of the unimolecular reactions. The nature of the dissociative state (S_0 or T_1) is often unknown and is assigned by conjecture and circumstantial evidence. Only in Bu^{t}NO and CF$_2$ClNO has the triplet state been directly implicated. No such data exist for HNO, NCNO, and CF$_3$NO. In addition, a complete set of unimolecular reaction rates that can be compared with statistical theories is still unavailable. Recent picosecond experiments of Knee *et al.*,[60] which probe the time-resolved appearance of products, should provide new clues regarding the effects of interstate couplings and incomplete energy randomization in small nitroso molecules.

3   Photodissociation of Aliphatic Nitrites – RONO

Spectroscopy

In contrast to the nitroso compounds whose dissociation can be described using statistical theories, the photodissociation of aliphatic nitrites is controlled by dynamical effects. Dissociation is fast (less than a rotational period), and product quantum-state distributions carry a memory of the initial excitation. There are several features that make the nitrites excellent candidates for photodissociation studies. Like aliphatic nitroso compounds, they have very similar electronic spectra and share a common dissociation mechanism. Thus, different molecules can be chosen to demonstrate specific features of a proposed mechanism. In addition, with NO as the photofragment, vector as well as scalar properties can be probed in unprecedented detail, giving a rather complete picture of the dissociation mechanism. Here, we will concentrate mainly on the 300—400 nm photodissociation of the following molecules: HONO,[68,69] CH$_3$ONO,[70–75] (CH$_3$)$_3$CONO (t-butyl nitrite, hereafter referred to as Bu^{t}ONO),[76] and (CH$_3$)$_2$NNO (dimethyl-nitrosamine, hereafter referred to as DMN).[77–79] The latter shows very similar spectroscopic and photodissociation properties and is therefore included. These molecules all exhibit very similar spectra and dissociation features and therefore will not be discussed separately. Rather, the different scalar and vector properties of their dissociation will be scrutinized, and general conclusions regarding the photodissociation mechanism will be derived. We will not discuss dissociation at other wavelengths,[74,75,79,80] since the results are not as yet complete, nor will we describe the infrared multiple-photon dissociation of CH$_3$ONO.[81,82]

[68] R. Vasudev, R. N. Zare, and R. N. Dixon, *Chem. Phys. Lett.*, 1983, **96**, 399.
[69] R. Vasudev, R. N. Zare, and R. N. Dixon, *J. Chem. Phys.*, 1984, **80**, 4863.
[70] F. Lahmani, C. Lardeux, and D. Solgadi, *Chem. Phys. Lett.*, 1983, **102**, 523.
[71] O. Benoist d'Azy, F. Lahmani, C. Lardeux, and D. Solgadi, *Chem. Phys.*, 1985, **94**, 247.
[72] F. Lahmani, C. Lardeux, and D. Solgadi, *Chem. Phys. Lett.*, 1986, **129**, 24.
[73] U. Brühlmann, M. Dubs, and J. R. Huber, *J. Chem. Phys.*, 1987, **86**, 1249.
[74] B. A. Keller, P. Felder, and J. R. Huber, *Chem. Phys. Lett.*, 1986, **124**, 135.
[75] B. A. Keller, P. Felder, and J. R. Huber, *J. Phys. Chem.*, 1987, **91**, 1114.
[76] D. Schwartz-Lavi, I. Bar, and S. Rosenwaks, *Chem. Phys. Lett.*, 1986, **128**, 123.
[77] M. Dubs and J. R. Huber, *Chem. Phys. Lett.*, 1984, **108**, 123.
[78] M. Dubs, U. Brühlmann, and J. R. Huber, *J. Chem. Phys.*, 1986, **84**, 3106.
[79] R. Lavi, I. Bar, and S. Rosenwaks, *J. Chem. Phys.*, in press.
[80] R. D. Kenner, F. Rohrer, and F. Stuhl, *J. Phys. Chem.*, 1986, **90**, 2635.
[81] D. S. King and J. C. Stephenson, *Chem. Phys. Lett.*, 1985, **114**, 461.
[82] D. S. King and J. C. Stephenson, *J. Chem. Phys.*, 1985, **82**, 2236.

The first absorption band of HONO and its organic esters (RONO) is the $\tilde{A}^1 A'' \leftarrow \tilde{X}^1 A'$ ($\pi^* \leftarrow n$) transition, which is localized mainly on the external O atom of the NO moiety.[28,83,84] The transition moment is perpendicular to the plane of the molecule and the major geometrical change upon excitation is lengthening of the NO bond, giving rise to a prominent Franck–Condon progression in the NO stretch. The most stable isomer of HONO is *trans*,[68,69] whereas for RONO it is *cis*[85] (the *cis/trans* ratio is 1.75 at 300 K).

Spectra in the 300—400 nm region consist of a series of prominent bands with well developed vibrational structure but diffuse band contours with no discernible rotational structure.[83] This indicates that the lifetime of the excited state is shorter than a rotational period but is sufficiently long to allow oscillation in the NO co-ordinate before dissociation. Dissociation gives rise to OH/RO and NO fragments in their ground electronic states.[28]

Below, we shall summarize experiments that probe the quantum-state distributions in OH($X^2\Pi$) in the case of HONO and NO($X^2\Pi$) in the case of RONO and R_2NNO. In the case of HONO it is very difficult to measure the NO distributions, because HONO exists in equilibrium with NO and NO_2. However, it may prove possible to monitor the NO fragment without interference in experiments with jet-cooled samples. Since the NO photofragment appears to be rotationally 'hot' (*vide infra*), it could be distinguished from NO contaminations that would be in only low rotational states following jet expansion, by using 'ON/OFF' experiments.[61] In the case of RONO, it is possible in principle to measure the RO distributions via LIF,[86–88] but so far such measurements have not been reported. NO has been probed by both one-[76,79] and two-photon LIF.[70–73,77,78] The experiments are typically done in the pump/probe mode, and the delay between the photolysis and detection lasers is usually kept to less than 50 ns. Most experiments were performed with 300 K samples, and the pressure was kept low enough to ensure collision-free conditions. Time-of-flight (TOF) measurements of the photofragments were carried out with jet-cooled samples of CH_3ONO.[74,75]

Vibrational- and Rotational-Energy Distributions

V,R energy distributions in OH($X^2\Pi$) from the photodissociation of HONO[68,69] and NO from CH_3ONO,[70,71,73] Bu^{t}NO,[76] and DMN[77,78] were reported. One should be aware that in fast photofragmentation processes the rotational distributions cannot be inferred straightforwardly from the raw experimental line intensities owing to alignment of the fragments[69,78] (see Chapter 4). The rotational populations of OH from HONO[69] and of NO

[83] G. W. King and D. Moule, *Can. J. Chem.*, 1962, **40**, 2057.
[84] C. Larrien, A. Dargelos, and M. Chaillet, *Chem. Phys. Lett.*, 1982, **91**, 465.
[85] P. Tarte, *J. Chem. Phys.*, 1952, **20**, 1570.
[86] G. Inoue, H. Akimoto, and M. Okuda, *Chem. Phys. Lett.*, 1979, **63**, 213.
[87] G. Inoue, M. Okuda, and H. Akimoto, *J. Chem. Phys.*, 1981, **75**, 2060.
[88] K. Fuke, K. Ozawa, and K. Kaya, *Chem. Phys. Lett.*, 1986, **126**, 119.

from DMN^{78} and CH_3ONO^{73} were derived with proper account for alignment effects. However, the rotational distributions are so distinctive and non-statistical that even the lack of proper corrections would not affect the qualitative conclusions.

$OH(X^2\Pi)$ rotational distributions were measured following photolysis of HONO at 369, 355, and 342 nm.[68,69] The populations in the $^2\Pi_{\frac{1}{2}}$ and $^2\Pi_{\frac{3}{2}}$ states can be characterized by temperatures: 385 ± 8 K and 240 ± 16 K for 369 nm photolysis, 364 ± 20 K and 253 ± 5 K for 355 nm photolysis, and 370 ± 10 K and 269 ± 17 K for 342 nm photolysis.[69] The results show that (i) the two spin–orbit states (F_1 and F_2) are not in equilibrium with each other, although the members of each of these are reasonably well characterized by a temperature, (ii) the F_1/F_2 population ratios are N-dependent and larger than statistical, and (iii) the rotational-state distributions within the F_1 and F_2 components are not very sensitive to changes in the number of quanta of v_2 excited in the HONO $\tilde{A}$ state.

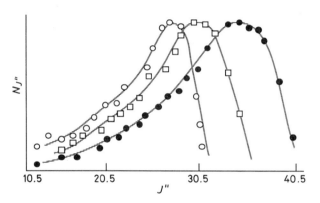

Figure 12 *Relative rotational populations of* $NO(X^2\Pi)$ *[*$v'' = 1$ (●)*,* $v'' = 2$ (□)*, and* $v'' = 3$ (○)*] following photodissociation of* CH_3ONO *at* 338 nm[71]
(Reproduced with permission from *Chem. Phys.*, 1985, **94**, 247)

Rotational distributions in $NO(X^2\Pi)$ (uncorrected for alignment effects) were reported for photolysis of CH_3ONO at 364, 350, 337, 327, and 318 nm using two-photon LIF.[70,71] In addition, NO distributions corrected for alignment were reported for 363.9 nm photolysis.[73] Typical rotational distributions for $NO(v'' = 1\text{—}3)$ are shown in Figure 12 for excitation at 338 nm. At this photolysis wavelength $NO(v'' = 0)$ is not populated (see below). Clearly, the distributions are not thermal and peak at high J''. The maximum population is shifted towards lower J'' as the quantum number v'' is increased, peaking at $J'' = 34.5$, 30.5, and 28.5 for $v'' = 1$, 2, and 3, respectively (Figure 12).[71] The rotational distributions within each $NO(v'')$ are independent of dissociation energy. Results obtained at 363.9 nm by a different group show a similarly 'hot' distribution.[73] A 'hot' rotational distribution has also been obtained for DMN at 363.5 nm[77,78] Results

Table 3 CH$_3$ONO *photolysis*: NO($X^2\Pi$) *vibrational distribution as a function of dissociative wavelength*[a]

$\lambda_{exc.}$[b]/nm	$E_{avl.}$/cm^{-1}	$N_{v''=0}$	$N_{v''=1}$	$N_{v''=2}$	$N_{v''=3}$	$N_{v''=4}$	$E_{vib.}^{CH_3ONO^*}$[c]/cm^{-1}	$\langle E_{vib.}^{NO}\rangle$/cm^{-1}
364 (3_0^1)	12 995	0.75	0.25	—	—	—	1175	470
355	13 650	0.25	0.50	0.20	0.05	—	1870	1960
350 (3_0^2)	14 055	0.20	0.50	0.30	<0.05	—	2275	2060
338 (3_0^3)	15 070	<0.05	0.375	0.425	0.20	—	3290	3400
327 (3_0^4)	16 065	—	0.15	0.33	0.52	—	4285	4400
318 (3_0^5)	16 930	—	—	0.13	0.55	0.32	5150	5880

[a] O. Benoist d'Azy, F. Lahmani, C. Lardeux, and D. Solgadi, *Chem. Phys.*, 1985, **94**, 247. [b] Parentheses indicate the vibrational transition. [c] $E_{vib.}^{CH_3ONO^*} = h\nu_{exc.} - E_{00}$ is the vibrational energy in the ν_3 stretching mode of the CH$_3$ONO $n\pi^*$ state

obtained with[78] and without[77] correcting for molecular alignment are qualitatively similar. The rotational distribution is non-thermal, peaking at $J'' = 25.5$ for $v'' = 0$. The average energies in rotation are 1515, 1340, and 1170 cm^{-1} for $v'' = 0$, 1, and 2, respectively.[78] Likewise, preliminary experiments on the 381.1 nm photolysis of Bu^{t}ONO yield highly excited, inverted distributions.[76]

The most complete vibrational distributions have been reported for NO from CH$_3$ONO[71] and are summarized in Table 3 as a function of dissociation energy. Several conclusions are immediately apparent: (i) the vibrational energy in NO($X^2\Pi$) at each photolysis wavelength is distributed mainly among 2 or 3 adjacent levels, (ii) when higher members of the 3_0^n vibrational progression in CH$_3$ONO are excited, the vibrational distribution reaches a maximum at higher v'', and (iii) the total vibrational-energy content of the NO fragment is close to the vibrational energy deposited in the N═O stretching mode of the excited CH$_3$ONO state. The vibrational population ratios reported for 363.5 nm DMN photodissociation are: $v_0/v_1/v_2/v_3 = 0.65/0.29/0.055/\sim 0.002$.[77,78] Preliminary analyses of the Bu^{t}ONO results indicate that, as with the case of CH$_3$ONO, the vibrational-energy distribution is very sensitive to the number of quanta excited in the N═O stretching mode, with higher vibrational quanta excited at shorter photolysis wavelengths.[76] No vibrational excitation in the OH($X^2\Pi$) fragment has been observed in HONO photolysis.[69] Some of the results are summarized in Tables 4 and 5.

The origin of the fragment V,R excitations has been addressed by several authors.[68–71,78] Vasudev *et al.*[69] suggest that the OH product rotations from HONO originate in the parent bending-type vibrations: the in-plane H—O—N bend, v_3, and the out-of-plane torsion, v_6. They use a model similar to the one considered previously by Freed *et al.*[89,90] and propose

Table 4 *Fragment energy distributions following photodissociation of* CH$_3$ONO *at* 350 nm

	NO($X^2\Pi$)					CH$_3$O	
	$v'' = 0$	$v'' = 1$	$v'' = 2$	$\langle E \rangle$/ cm^{-1}	$\dfrac{\langle E \rangle}{E_{avl.}}$/%	$\langle E \rangle$/ cm^{-1}	$\dfrac{\langle E \rangle}{E_{avl.}}$/%
$E_R{}^a$	1900	1820	1350	1700^{a,b}	12	2300^c	16
E_V	0	1876	3720	2060^{a,b}	15		
$E_T{}^d$				4070	29	3930	28
E_{total}				7830	56	6230	44

^{a} O. Benoist d'Azy, F. Lahmani, C. Lardeux, and D. Solgadi, *Chem. Phys.*, 1985, **94**, 247. ^{b} Calculated using $v'' : v'' : v'' = 0.20 : 0.50 : 0.30$. ^{c} Calculated using all other values and $E_{avl.} = 14\,055$ cm^{-1}. ^{d} B. A. Keller, P. Felder, and J. R. Huber, *Chem. Phys. Lett.*, 1986, **124**, 135

[89] K. F. Freed, M. D. Morse, and Y. B. Band, *Faraday Discuss. Chem. Soc.*, 1979, **67**, 297.
[90] M. D. Morse and K. F. Freed, *J. Chem. Phys.*, 1983, **78**, 6045.

Table 5 *Fragment energy distributions following photodissociation of* DMN *at 363.5 nm[a]*

	$NO(X^2\Pi)$					$(CH_3)_2N$	
	$v'' = 0$	$v'' = 1$	$v'' = 2$	$\frac{<E>/}{cm^{-1}}$	$\frac{<E>}{E_{avl.}}/\%$	$\frac{<E>/}{cm^{-1}}$	$\frac{<E>}{E_{avl.}}/\%$
E_R	1515	1340	1170	1440^b	10	5200^c	36
E_V	0	1876	3720	760^b	5		
E_T				4190	29	2860	20
E_E				60	0.4		
E_{total}				6450	44	8060	56

[a] From M. Dubs, U. Brühlmann, and J. R. Huber, *J. Chem. Phys.*, 1986, **84**, 3106. [b] Assuming $v''_0 : v''_1 : v''_2 = 0.65 : 0.29 : 0.055$. [c] Calculated from all other values and $E_{avl.} = 14\,510\,cm^{-1}$

that the higher OH product rotations arise predominantly from v_3, while the lower fragment rotations originate from both v_3 and v_6.

Another possible source of fragment rotation involves the impulse associated with the dissociation.[68,69] In this model,[91] fragment rotation arises from the torque exerted on the fragments along the bond that breaks. It is assumed that each fragment is rigid and the available energy is transferred to motion along the breaking bond. The available energy is distributed between fragment rotation and translation. Vasudev *et al.*[69] find that the OH rotational distribution from HONO cannot be matched by such a model, presumably because the OH bond is not 'rigid' enough for the H atom to respond to the impulse exerted on the O atom during bond cleavage. Benoist d'Azy *et al.*[71] demonstrate that a modified impulsive model can explain the NO rotational distributions from CH_3ONO. These authors observe an amount of vibrational energy in the NO photofragment that is almost equal to the energy content in the $\tilde{A}^1A''$ 3_0^n vibrational level. Since this energy is apparently unavailable for R,T excitation, the available energy has been taken as:

$$E'_{avl.}(v'') = E_{00} - D_0 - hv_{v''}^{NO} \tag{6}$$

where $hv_{v''}^{NO}$ is the energy of the NO vibrational level and E_{00} is the band origin of the $\tilde{A}^1A''$ state. Inserting the experimental value of $E'_{avl.}(v'')$ into the impulsive model, a much better fit to the experimental $<E_{rot.}>$ is obtained. A similar model would probably explain the rotational distributions in the photodissociation of other nitrites as well, and was also used successfully to fit the NO translational-energy distribution in CH_3ONO.[74,75]

Benoist d'Azy *et al.* tried several models in order to explain the observed vibrational-energy distributions in the NO photofragment from CH_3ONO.[71] Following Vasudev *et al.*,[69] they used a Franck–Condon model assuming that, since the $\tilde{A}^1A''$ state vibrates before dissociation, all memory of initial excitation has been lost and the $\tilde{A}^1A''$ state reaches its

[91] A. F. Tuck, *J. Chem. Soc., Faraday Trans. 2*, 1977, **73**, 689.
[92] R. T. Pack, *J. Chem. Phys.*, 1976, **65**, 4765.
[93] L. Werner, B. Wunderer, and H. Walther, *Chem. Phys.*, 1981, **60**, 109.

equilibrium position along the NO co-ordinate. They found that this model does not fit the observed distributions. They then turned to a different model,[92] used previously to explain the photofragmentation of NOCl.[93] This model assumes that the excited potential surface of CH_3ONO can be described in terms of two independent co-ordinates: the dissociation co-ordinate along the O—N bond and the terminal N=O bond. The surface is repulsive in the dissociation co-ordinate but still bonding with respect to the terminal N=O bond. The ground electronic state is represented by a two-dimensional harmonic oscillator along the same co-ordinates, and the probability distribution of motions in the excited state can be obtained using the Franck–Condon principle. The model also assumes that the vibrational energy deposited in the v_3 mode of CH_3ONO remains in the NO fragment, as observed experimentally. Reasonable agreement is obtained with the experimental observations for a potential surface with a rather smooth repulsive potential along the dissociative co-ordinate.[71] This model tends to favour direct dissociation from the excited surface over a mechanism that involves predissociation following crossing to a repulsive electronic surface. Very recent MCSCF calculations yield a qualitatively similar excited potential-energy surface.[94]

Translational-Energy Distributions

Fragment translational energies have been determined in two ways: from Doppler profiles of the individual rotational lines of $OH^{68,69}$ and NO^{78} and from TOF and angular distribution measurements of photofragments using mass-resolved ionization detection.[74,75,91]

For the case of CH_3ONO photolysed at 350 nm, the centre-of-mass (CM) kinetic energy was derived from TOF spectra.[74,75] The molecular-beam apparatus used a pulsed molecular beam and mass spectrometric detection of the fragments. TOF spectra of both NO and CH_3O were recorded, and these were the sole dissociation products. The translational-energy distribution is somewhat asymmetric, with $<E_{trans.}> = 8000\ cm^{-1}$ and a width of $2100\ cm^{-1}$ ($<E_{trans.}^{NO}> = 4070 \pm 180\ cm^{-1}$, $<E_{trans.}^{CH3O}> = 3930 \pm 170\ cm^{-1}$). For the case of C_2H_5ONO, TOF spectra were obtained following 347 nm photolysis.[91] $<E_{trans.}^{NO}> = 6000\ cm^{-1}$ was obtained, with $<E_{trans.}^{NO}> = 3600\ cm^{-1}$ and $<E_{trans.}^{C2H5NO}> = 2400\ cm^{-1}$. The translational-energy distribution could be modelled using the modified impulsive model as described above for NO rotations.[91]

Photofragment translational energies for the photodissociation of $HONO$,[68,69] CH_3ONO,[73] and DMN^{78} were derived from the LIF Doppler profiles. In the absence of correlations between fragment v and J (*vide infra*), the CM angular distribution for a single-fragment recoil velocity and for a linearly polarized photolysis laser is given by:[69]

$$w(\theta) = \frac{1}{4\pi}[1 + \beta P_2(\cos\theta)] \tag{7}$$

[94] M. Nonella and J. R. Huber, *Chem. Phys. Lett.*, 1986, **131**, 376.

where P_2 is the second Legendre polynomial and θ is the angle between the fragment recoil direction and the photolysis laser electric vector (ε_p). The anisotropy parameter β lies between $+2$ and -1, corresponding, respectively, to the limiting cases of recoil along and perpendicular to the parent molecule transition moment. The corresponding fragment Doppler profile $D(\Delta v)$ is given by:[69]

$$D(\Delta v_0) = \frac{1}{\Delta v_D} [1 + \beta P_2(\cos\theta) P_2(\Delta v_0/\Delta v_D)] \tag{8}$$

where $\Delta v_D = v_0(v/c)$ is the maximum Doppler shift and Δv_0 is the displacement from the line centre v_0. The Doppler profile for a distribution of recoil velocities has to be obtained by convolution with the velocity distribution function. In the case when J and v are correlated, different spectroscopic branches may give different profiles, and the deconvolution of the Doppler lineshapes becomes quite complex[78,95-97] (see Chapter 4). Values of β derived for the nitrites from Doppler profiles[69,73,78] as well as from angular distributions of the photofragments[74,75,91] are summarized in Table 6.

Table 6 *Alignment* $[\mathscr{A}_0^{(2)}]$ *and recoil anisotropy* (β) *parameters*

Molecule	λ/nm	J values[a]	$\mathscr{A}_0^{(2)}$	β
HONO	369, 342, 355	0.5—5.5	0.1—0.3[b]	−0.8[b]
CH_3ONO	355,[c] 350[d]	20.5—40.5	0.5[c]	−0.7[d]
C_2H_5ONO	347	15.5—40.5	0.38[e]	−0.7[f]
$(CH_3)_3CONO$	381.1	30.5—45.5	0.32[g]	
$(CH_3)_2NNO$	363.5	10.5—40.5	0.1[e]	−0.56[e]
		25.5—45.5	0.18[h]	

[a] Used in determining $\mathscr{A}_0^{(2)}$. [b] R. Vasudev, R. N. Zare, and R. N. Dixon, *J. Chem. Phys.*, 1984, **80**, 4863. [c] F. Lahmani, C. Lardeux, and D. Solgadi, *Chem. Phys. Lett.*, 1986, **129**, 24. [d] B. A. Keller, P. Felder, and J. R. Huber, *J. Phys. Chem.*, in press. [e] M. Dubs, U. Brühlmann, and J. R. Huber, *J. Chem. Phys.*, 1986, **84**, 3106. [f] A. F. Tuck, *J. Chem. Soc., Faraday Trans.*, 1977, **2**, 689. [g] D. Schwartz-Lavi, I. Bar, and S. Rosenwaks, *Chem. Phys. Lett.*, 1986, **128**, 123. [h] R. Lavi, I. Bar, and S. Rosenwaks, *J. Chem. Phys.*, in press

The OH Doppler lineshapes from HONO photolysis are shown in Figure 13; for comparison, the profile of 300 K OH is also shown.[69] The spectrum was obtained with two parallel counterpropagating laser beams using 355 nm photolysis and $\varepsilon_p \| \varepsilon_d$, where ε_d is the polarization vector of the detection laser. The spectra show the $P_2(5)$ and $^PQ_{12}(5)$ lines of the OH($A-X$) (1, 0) transition. The observed lineshape corresponds to an anisotropic OH recoil distribution typical of a perpendicular transition.[69] Under the reported experimental conditions, correlations between J and v could not

[95] R. N. Dixon, *J. Chem. Phys.*, 1986, **85**, 1866.
[96] G. E. Hall, S. Sivakumar, P. L. Houston, and I. Burak, *Phys. Rev. Lett.*, 1986, **56**, 1671.
[97] G. E. Hall, N. Sivakumar, R. Ogorzalek, G. Chawla, H.-P. Haerri, P. L. Houston, I. Burak, and J. W. Hepburn, *Faraday Discuss. Chem. Soc.*, 1986, **82**, in press.

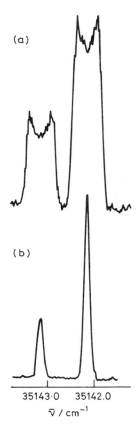

(a)

(b)

35143·0　　35142.0

$\bar{\nu}$ / cm^{-1}

Figure 13 *Narrow-band excitation spectra in the* 35 141—35 146 cm^{-1} *region showing the* $P_2(5)$ *and* $^PQ_{12}(5)$ *lines of* (a) *the* OH *fragment produced by photolysis of* HONO *at* 355 nm *with* $\varepsilon_p \parallel \varepsilon_d$ *and* (b) *equilibrated* (300 K) OH[69]
(Reproduced with permission from *J. Chem. Phys.*, 1984, **80**, 4863)

have been discerned.[95] In the analysis, the assumption has been made of a very narrow (delta function) distribution of fragment velocities, and the best results were obtained with $\beta = -0.8$. This value is close to the limiting value of -1 for a perpendicular transition and implies a dissociation rate much faster than the rotational period.[69,98]

In a similar way, Dubs *et al.* analysed the lineshapes obtained in the two-photon LIF detection of NO from DMN, using the $A^2\Sigma^+ \leftarrow X^2\Pi$ transition.[78] These authors treat in detail the determination of the velocity distribution in the presence of rotational alignment. They show that, because of the alignment of the rotational angular momentum, slightly different Doppler profiles are observed when probing R- and S-branch lines that originate from the same level (Figure 14). If no correction is made for these alignment effects, different $<E_{\text{trans.}}>$ are obtained for the R- and S-branch

[98] R. N. Zare and D. R. Herschbach, *Proc. IEEE*, 1963, **51**, 173.

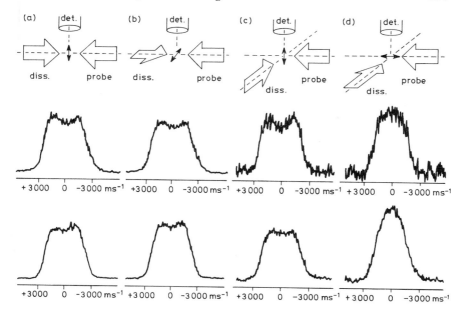

Figure 14 *Doppler profiles of* NO *from the 363.5 nm DMN photolysis obtained using four different pump/probe geometries.*[78] *The orientation of the probe laser polarization vector is given by the two-directional arrow. Upper row: profiles of the $R_{12}(J'' = 26.5)$ line of the $A \leftarrow X$ (0,0) transition; lower row: as above but for the $S_{22}(J'' = 26.5)$ transition. The abscissa scale indicates the Doppler velocity v_k/ms^{-1}*

(Reproduced with permission from *J. Chem. Phys.*, 1986, **84**, 3106)

lines.[78] The results of all the studies show that in the photodissociation of aliphatic nitrites a major fraction of the available energy is channelled into fragment translation, in agreement with the modified impulsive model described above (see Tables 4 and 5).

Rotational Alignment of the Fragments

When a molecule is excited using polarized radiation, the optical transition moment, $\boldsymbol{\mu}$, is preferentially aligned along the electric vector of the excitation (ε_p). This spatial alignment may result in anisotropic distributions of the fragment $\boldsymbol{J}$ with respect to $\boldsymbol{\mu}$. In the RONO molecules the $\tilde{A}$–$\tilde{X}$ transition moment is perpendicular to the molecular plane, and thus the photoexcited molecules are preferentially in a plane perpendicular to ε_p. When the dissociation time is shorter than the parent rotational period, the fragments will show the same spatial anisotropies as the excited parent molecule. Thus, if dissociation is planar, and the fragments rotate in the plane, J'' is expected to show the same spatial distribution as $\boldsymbol{\mu}$ relative to ε_p. This alignment can be probed using LIF, since transition moments for the P- and R-branches of NO or OH are in the plane of rotation, while for the Q-branch they lie parallel to the rotational angular momentum. It has been shown that the relative efficiency of detection of P/R- *vs.* Q-branch lines depends on the relative orientation of ε_p and ε_d.[99]

Spatial angular momentum anisotropy in photodissociation has been treated in detail by Greene and Zare.[99] It can be expressed in terms of the alignment parameter $\mathscr{A}_0^{(2)}$, the initial-state expectation value of the quadrupole moment of the total angular momentum operator J. For a perpendicular transition and planar dissociation, the upper (high-J) limit for $\mathscr{A}_0^{(2)}$ is 0.8, corresponding to the case of preferential population of the Zeeman levels with $|M_J| \sim J$. By observing the relative P/R- *vs.* Q-branch line intensities as a function of laser polarization, the degree of alignment can be estimated. The relevant formulae for two-photon LIF excitation were derived by Dubs *et al.*[78] The alignment factors obtained for HONO,[69] CH_3ONO,[72] C_2H_5ONO,[78] DMN,[78,79] and Bu^tONO[76] photolyses are summarized in Table 6, and results obtained using one-[79] and two-photon LIF[78] are similar. In HONO photolysis the OH $\mathscr{A}_0^{(2)}$ values are N-dependent, ranging from 0.1 for $N = 1$ to 0.3 for $N = 5$. However, the $\mathscr{A}_0^{(2)}$ factors for all N values are positive and tend towards a parallel orientation of the rotational angular momentum with respect to the transition moment at high N. Positive limiting values of $\mathscr{A}_0^{(2)}$ are also obtained for NO in the photolysis of other RONO molecules.

The major conclusions to be drawn from the fragment alignment measurements are: (i) all values are positive, indicating $J \parallel \mu$, and (ii) the value of $\mathscr{A}_0^{(2)}$ decreases with the size and complexity of the molecule. Thus, the most likely photodissociation mechanism involves excitation via a perpendicular transition followed by fast dissociation from a planar transition state. Deviations from the limiting alignment value can be due to the rate of dissociation, which allows parent rotation prior to bond cleavage. In those cases where comparisons between recoil anisotropies and rotational alignments are possible (Table 6), both correspond to excited-state lifetimes of less than 1 ps, shorter than a rotational period.

Λ-*Doublet Populations*

The preferential population of one of the Λ-doublets of a $^2\Pi$ fragment indicates a preferential orientation of the single-electron $p\pi$ orbital lobes of the fragment relative to its plane of rotation. Several publications have dealt recently with this phenomenon, and most of the concepts have now been clarified.[100-102] We note, however, that some confusion still exists regarding the assignment of the total electronic parity of the Π state, which determines its rotational symmetry.[69,100,101] For example, the Π^- orbitals

[99] C. H. Greene and R. N. Zare, *Annu. Rev. Phys. Chem.*, 1982, **33**, 119.
[100] P. Andresen and E. W. Rothe, *J. Chem. Phys.*, 1985, **82**, 3634.
[101] M. H. Alexander and P. J. Dagdigian, *J. Chem. Phys.*, 1984, **80**, 4863.
[102] W. D. Gwinn, B. E. Turner, W. M. Goss, and G. L. Blackman, *Astrophys. J.*, 1973, **179**, 789.

in references 69, 72, and 76 are defined as Π^+ in ref. 78. The origin of this disagreement lies in the different definitions of the parity used by different authors as a result of treating different Hund's coupling cases.[69,100] There is, however, an agreement regarding the geometrical interpretation of the relative intensities of the different branch lines of the probed transition. In one-photon LIF, Q-branch lines probe the $p\pi$ orbital perpendicular to the plane of rotation, while P- and R-branch lines probe $p\pi$ orbitals that are in the plane of rotation. The correlation between the branches probed in two-photon LIF and the orientation of the $p\pi$ lobes is discussed by Lahmani *et al.*[72] and Dubs *et al.*[78] Therefore, to avoid further confusion, we will refer to the $p\pi$ orbitals that are perpendicular or parallel to the plane of rotation of the fragment as $\Pi_{\aleph}$ and Π_{∞}, respectively. It should be noted that for small J the π lobes are less well oriented,[100–102] and the degree of electron alignment as a function of J is determined by $\lambda = A/B$, where A is the spin–rotation splitting and B is the rotational constant ($\lambda = -7.5$ for OH and 72.6 for NO).[100]

Vasudev *et al.*[68,69] measured the relative Λ-doublet populations in the $X^2\Pi$ state of OH generated by HONO photolysis and found that $\Pi_{\infty} > \Pi_{\aleph}$, showing a preferential orientation of the π lobe in the plane of rotation. Thus the OH $p\pi$ orbital, which is directed towards the NO fragment in the ground state, does not change its orientation during excitation and fragmentation, indicating that dissociation is fast and takes place from a planar transition state.

In the case of the NO fragment all studies (CH_3ONO,[72,73] DMN,[78,79] and Bu^tONO[76]) indicate that $\Pi_{\aleph} > \Pi_{\infty}$. This is not surprising in view of the symmetry change upon excitation ($\tilde{A}^1A'' \leftarrow \tilde{X}^1A'$). The excited planar RONO surface is of A'' symmetry, and its electronic configuration can be described in a simplified way by the promotion of a non-bonding electron in the symmetric orbital localized in the N=O moiety of the $\tilde{X}^1A'$ state to the lowest π^* antibonding orbital which is of a'' symmetry. If dissociation is planar, the NO $p\pi$ electron lobe will retain this symmetry with respect to the plane of the molecule. Since $\Pi_{\aleph}$ is antisymmetric with respect to reflection in the plane of rotation, it will be preferentially populated in the limit of high J (where Hund's case b applies). In 355 nm CH_3ONO photolysis and 363.5 nm DMN photolysis the extent of π-orbital alignment is smaller than the calculated limiting value, indicating deviations from a completely planar dissociation. Such deviations may be caused by parent rotation, out-of-plane torsional vibrations, or incomplete localization of the charge distribution in the antibonding π^* orbital of the N=O group. Lahmani *et al.* obtain the same degree of localization for jet-cooled and 300 K samples, thus excluding parent rotation as a source of deviations.[72]

In summary, the Λ-doublet populations indicate that dissociation is fast and planar, with deviations from planarity that probably increase with the size and complexity of the parent molecule.

Spin–Orbit Populations

Both OH and NO have $^2\Pi_{\frac{1}{2}}$ and $^2\Pi_{\frac{3}{2}}$ spin–orbit states that have small energy separations.[69] Thus, their relative populations affect the total energy balance very little. However, a preferential population of one spin–orbit state may also mean a preferential orientation of the spin with respect to the electron orbital angular momentum. Such preferential population of the spin–orbit states has been found for OH in HONO photolysis, where the higher spin state is more populated than predicted by statistics.[68,69] In NO generated via CH_3ONO,[72] DMN,[78] and Bu^tONO[76] photolyses, the lower state is slightly more populated than the upper state. The reason for such preferential spin orientation is not clear, since for a fast dissociation from a singlet surface there is no preferred axis for the spin of the odd electron of each fragment. It has been suggested that the preferential spin–orbit population may result either from curve crossings or from exit-channel interactions leading to electronic energy transfer between the RO and NO photofragments.[69]

Correlations between J and v

Since both J and v are correlated with μ, one also expects to see correlations between J and v. Such correlations have recently been treated and demonstrated quantitatively for the case of fragments resulting from the photodissociation of triatomic molecules.[96,97] For example, if J is perpendicular to v, then molecules moving with v parallel to the direction of propagation of the probe laser will have only $M_J = 0$, while those moving with v perpendicular to the laser propagation direction will have a broad distribution of M_J values.[97] Since the fragment rotational alignment parameter, $\mathscr{A}_0^{(2)}$, is dependent on M_J, it will also depend on v, and will vary along the Doppler line profile.[78] As shown before, different values of $\mathscr{A}_0^{(2)}$ cause changes in the relative intensities of the different spectroscopic branches, and consequently one expects that the Doppler shapes for lines of different branches will be slightly different. Such differences in Doppler profiles were indeed observed in DMN photolysis between the R_{12} and S_{22} transitions of NO probed via two-photon LIF,[78] and these are shown in Figure 15. If the lineshapes are not corrected for alignment effects, different branches yield different translational energies. Although a quantitative analysis of the lineshapes is not possible for polyatomic molecules with a velocity distribution, qualitative considerations of the changes in $\mathscr{A}_0^{(2)}$ with v indicate a preferred perpendicular orientation of v with respect to J, thus supporting a dissociation mechanism that involves a planar transition state. Similar results were recently obtained for the 363.9 nm photodissociation of CH_3ONO.[73]

Summary

By combining the results of the different investigations described above, a fairly detailed picture of the dissociation mechanism of aliphatic nitrites emerges.

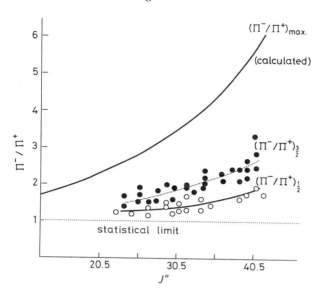

Figure 15 *Experimental and theoretical Λ-doublet population ratios in* $NO(X^2\Pi)$ *($v'' = 1$) produced by* 355 nm *photolysis of* CH_3ONO.[72] *Note that* Π^- *and* Π^+ *are referred to as* Π_8 *and* Π_∞, *respectively, in the text* (Reproduced with permission from *Chem. Phys. Lett.*, 1986, **129**, 24)

Upon excitation at 300—400 nm, the $\tilde{A}^1A''$ excited surface is reached. The $\tilde{A}^1A'' \leftarrow \tilde{X}^1A'$ ($\pi^* \leftarrow n$) transition moment is perpendicular to the plane of the molecule, and excitation involves promotion of a non-bonding electron localized on the N=O moiety to an antibonding orbital of a'' symmetry. The major geometrical change upon excitation is the lengthening of the NO bond. The excited-state potential surface is probably slightly bonding in the N=O co-ordinate, while it is rather steeply repulsive along the O—N co-ordinate. Consequently, the molecule can vibrate several times ($\tau_{vib.} \simeq 3 \times 10^{-14}$ s) before dissociation without significantly perturbing the motion along the repulsive co-ordinate. Such a surface has recently been obtained for CH_3ONO using MCSCF calculations.[94]

From the measured vector properties, it can be concluded that dissociation occurs from a largely planar transition state and the fragments rotate in the plane of the molecule (*i.e.* $J \parallel \mu$ and $J \perp v$). The lifetime of the excited state as derived from the recoil anisotropy parameter is <1 ps (≤0.2 ps for CH_3ONO[73] and HONO[69] and ~0.4 ps for DMN[78]) and is shorter than the rotational period (~10^{-12} s). The $p\pi$ lobe of the single-electron NO orbital is preferentially oriented perpendicular to the molecular plane, while the OH $p\pi$ lobe in HONO photodissociation is in the plane of rotation of OH (the plane of the excited parent state). A simplified picture of the dissociation geometry is presented in Figure 16, and deviations from the idealized picture increase, as expected, with the size and complexity of the molecule.

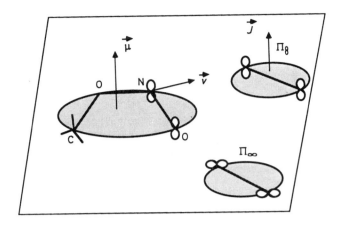

Figure 16 *An idealized planar fragmentation of* RONO *molecule.*[76] *The fragment rotational plane is identical to the parent molecule plane. The symbols are defined in the text*
(Adapted from *Chem. Phys. Lett.*, 1986, **128**, 123)

The derived scalar quantities $(E,V,R,T$ energy distributions) point to an impulsive dissociation mechanism. The shape of the excited potential surface leads to initial excitation of v_3 (N=O stretch), which tends to stay in the NO fragment, yielding increasingly 'hot' NO vibrational populations as the excitation wavenumber increases. The average energy in NO vibration is close to the initial energy in the v_3 parent vibration, and the distribution is rather narrow. The balance of the excess energy is deposited in R,T degrees of freedom, as predicted by an impulsive mechanism that assumes a rigid N=O bond. An exception to the impulsive model is OH rotation from HONO, probably because the OH bond is not rigid enough. The source of OH rotation is presumably in-plane vibrations and out-of-plane torsions of the excited HONO.

Acknowledgements. We wish to thank C. X. W. Qian and I. Nadler for many stimulating discussions concerning the predissociation of nitroso compounds. We are also grateful to S. Rosenwaks, J. Pfab, D. Solgadi, J. R. Huber, and J. Knee for communicating results to us before publication. Support from the National Science Foundation, the Air Force Office of Scientific Research, and the Army Research Office is gratefully acknowledged.

CHAPTER 6

The Dissociation Dynamics of Highly Vibrationally Excited Molecules

F. F. CRIM

1 Introduction

Highly vibrationally excited molecules play a central role in many electronic photodissociation and photochemical processes as well as presenting rich vibrational photodissociation dynamics of their own. Because internal conversion from an electronically excited state to high vibrational levels of the ground state is often the first process that occurs subsequent to electronic excitation, the resulting photochemistry is actually that of highly vibrationally excited molecules. In fact, it is possible to exploit this pathway as a preparation technique for highly vibrationally excited molecules for studies of collisional energy transfer[1,2] and unimolecular reactions,[3] and there are direct demonstrations of the role of vibrationally excited molecules formed by internal conversion in photochemical reactions.[4] Several state-resolved studies of the unimolecular decomposition of small molecules[5-8] have relied on internal conversion as a preparation scheme, and detailed measurements have shown internal conversion followed by vibrational predissociation to

[1] H. Hippler, L. Lindemann, and J. Troe, *J. Chem. Phys.*, 1985, **83**, 3906; J. E. Dove, H. Hippler, and J. Troe, *J. Chem. Phys.*, 1985, **82**, 1907 and references cited therein.
[2] M. J. Rossi, J. R. Pladzievicz, and J. R. Barker, *J. Chem. Phys.*, 1983, **78**, 6695; W. Forst and J. R. Barker, *J. Chem. Phys.*, 1985, **83**, 124 and references cited therein.
[3] H. Hippler, K. Luther, J. Troe, and H. J. Wendelken, *J. Chem. Phys.*, 1983, **79**, 239 and references cited therein.
[4] N. Nakashima, N. Shimo, N. Ikeda, and K. Yoshihara, *J. Chem. Phys.*, 1984, **81**, 3738; N. Ikeda, N. Nakasima, and K. Yoshihara, *J. Chem. Phys.*, 1985, **82**, 5285.
[5] I. Nadler, M. Noble, H. Reisler, and C. Wittig, *J. Chem. Phys.*, 1985, **82**, 2608; C. X. W. Qian, I. Nadler, M. Noble, H. Reisler, and C. Wittig, *J. Chem. Phys.*, 1985, **83**, 5573.
[6] D. J. Bamford, S. V. Filseth, M. F. Foltz, J. W. Hepburn, and C. B. Moore, *J. Chem. Phys.*, 1985, **82**, 3032.
[7] H. L. Dai, R. W. Field, and J. L. Kinsey, *J. Chem. Phys.*, 1985, **82**, 1606.
[8] D. J. Nesbitt, H. Petek, M. F. Foltz, D. J. Bamford, and C. B. Moore, *J. Chem. Phys.*, 1985, **82**, 223.

be the crucial steps in some photodissociation processes.[9] Highly vibrationally excited molecules are not only important in photodissociation and photochemistry. They are the reactive species in much thermal chemistry and are part of combustion processes and atmospheric chemistry.

One approach to isolating the dissociation dynamics of highly vibrationally excited molecules is to prepare them with large amounts of energy by single-photon excitation of overtone vibrations. This excitation process occurs solely on the ground electronic surface and, like electronic excitation, provides a known energy increment to the molecule. The state selectivity depends on the details of the couplings within the molecule and on the nature of the exciting light, but vibrational overtone excitation is potentially highly selective. It is likely to prepare a different initial distribution of internal energy than electronic excitation followed by internal conversion, where the internal-conversion process determines the vibrational states, and offers the opportunity to probe dissociation dynamics of systems that are initially far from microcanonical equilibrium. Vibrational overtone excitation can also serve as the first step in a two-photon excitation process that reaches regions of an electronically excited potential-energy surface that are inaccessible to conventional one-photon excitation. Such vibrationally mediated electronic photodissociation is a means of obtaining vibrational overtone spectra for bound molecules and of exploring unique regions of the excited-state potential-energy surface.

The questions that are common to electronic and vibrational photodissociation concern the initial preparation and the subsequent behaviour of the molecule. What is excited? What happens to the excited molecule? We address these questions experimentally by obtaining *vibrational overtone spectra*, by measuring *dissociation rates* of highly vibrationally excited molecules, and by determining *product quantum-state distributions* in reactions initiated by vibrational overtone excitation. Analysing vibrational overtone excitation spectra is a means of identifying the state initially prepared by the excitation laser or, equivalently, the couplings that are operative in the molecule. This approach is most fruitful in small molecules, where relatively few states are involved in the dynamics, but it is also quite informative for large molecules. An important feature of our spectroscopic studies and rate measurements is the use of molecules cooled in a free-jet expansion. Cooling the molecules reduces the congestion in the spectra and the distribution of initial energies in the rate measurements, both of which arise from the thermal population of a large number of vibrational and rotational states in a room-temperature sample. We obtain dissociation rates both directly, by observing the temporal evolution of the products, and indirectly, by analysing the widths of individual spectral features. Product-state distributions and rate measurements provide an incisive comparison with theoretical predictions, particularly near the reaction threshold energy.

[9] R. D. Bower, R. W. Jones, and P. L. Houston, *J. Chem. Phys.*, 1983, **79**, 2799.

2 Spectroscopy

Vibrational Overtone Spectroscopic Models

Many aspects of vibrational overtone spectroscopy are consistent with a local-mode model[10,11] in which the vibrational state that has a significant optical transition probability from the ground state corresponds to the stretching of the bond to a light atom. The coarse pattern of the vibrational transition frequencies follows that of an anharmonic oscillator with a mechanical frequency and anharmonicity characteristic of the type of bond, an OH or CH bond for example, and of its environment in the molecule. This extremely useful and often predictive picture is only the simplest description of the excitation in the molecule. A more detailed model uses the stretching state as the zero-order state that carries the oscillator strength of the transition and considers its coupling to other states in the molecule.[12-14] This model is strongly reminiscent of radiationless transition theory[15] and usually divides the background states into a relatively small group of states that are strongly coupled to the zero-order state and a much larger group of weakly coupled states. The utility of this basic picture is clear,[12-14] but the challenge is the identification of the strongly coupled states.

Vibrations that lie near the zero-order state in energy and possess significant kinetic or potential coupling to it are the most likely candidates for strongly coupled states. One type of state that has proven to be important in understanding the vibrational overtone spectroscopy of several molecules[16-22] is the background state that is in Fermi resonance with the zero-order stretching state. For example, Sibert *et al.*[16] suggest that the CH stretch/CCH bend combinations that are in Fermi resonance with the pure CH stretching state are the strongly coupled states in benzene, and they find that coupling of these to tiers of background states reproduces the substantial ($\sim 100\,\text{cm}^{-1}$) linewidths observed experimentally.[23] An important notion in this Fermi resonance model is the 'detuning' of the resonance between the zero-order (CH stretching) state and the Fermi resonant stretch/bend combinations by the anharmonicity of the CH stretching vibration.

[10] B. Timm and R. Mecke, *Z. Phys.*, 1936, **98**, 363; R. Mecke, *Z. Elektrochem.*, 1950, **38**, 1950.
[11] B. R. Henry, *Acc. Chem. Res.*, 1977, **10**, 207; *Vib. Spectra Struct.*, 1981, **10**, 269; L. Halonen and M. S. Child, *Adv. Chem. Phys.*, 1984, **57**, 1; M. S. Child, *Acc. Chem. Res.*, 1985, **45**, 45.
[12] P. R. Stannard and W. M. Gelbart, *J. Phys. Chem.*, 1981, **85**, 1981.
[13] M. L. Sage and J. Jortner, *Adv. Chem. Phys.*, 1981, **47**, 293.
[14] D. F. Heller and S. Mukamel, *J. Chem. Phys.*, 1979, **70**, 463; 'Picosecond Phenomena', ed. C. V. Shank, E. P. Ippen, and S. L. Shapiro, Springer, New York, 1978, p. 51.
[15] P. Avouris, W. M. Gelbart, and M. A. El-Sayed, *Chem. Rev.*, 1977, **77**, 793.
[16] E. L. Sibert, tert., W. P. Reinhardt, and J. T. Hynes, *J. Chem. Phys.*, 1984, **81**, 1115.
[17] J. W. Perry, D. J. Moll, A. Kuppermann, and A. H. Zewail, *J. Chem. Phys.*, 1985, **82**, 1195.
[18] G. J. Scherer, K. K. Lehmann, and W. Klemperer, *J. Chem. Phys.*, 1985, **81**, 5319.
[19] H.-R. Dübal and M. Quack, *J. Chem. Phys.*, 1985, **81**, 3779.
[20] J. E. Baggott, M.-C. Chuang, R. N. Zare, H.-R. Dübal, and M. Quack, *J. Chem. Phys.*, 1985, **82**, 1186.
[21] S. Peyerimhoff, M. Lewerenz, and M. Quack, *Chem. Phys. Lett.*, 1984, **109**, 563.
[22] K. von Puttkamer, H.-R. Dübal, and M. Quack, *Faraday Discuss. Chem. Soc.*, 1983, **75**, 263.
[23] K. V. Reddy, D. F. Heller, and M. J. Berry, *J. Chem. Phys.*, 1982, **76**, 2814.

There is no Fermi resonance for the first vibrational state of the CH stretch since its frequency of $3157 \, cm^{-1}$ is substantially more than twice the CCH bending frequency of $\sim 1300 \, cm^{-1}$. However, the anharmonicity of the CH stretching vibration brings its fifth overtone, $6\nu_{CH}$, close in energy to the combinations $5\nu_{CH} + 2\nu_{CCH}$, $4\nu_{CH} + 4\nu_{CCH}$, *etc.* Thus, Sibert *et al.*[16] predict strong interactions in this region that diminish for both higher and lower levels of excitation.

Fermi resonance interaction models are quite clearly applicable to the vibrational overtone spectra of several other molecules. Perry *et al.*[17] show that a two- or three-state model using the Fermi resonance between the CH stretch and degenerate bending modes in CD_3H describes much of the structure in the fifth vibrational overtone spectrum. Scherer *et al.*[18] find the same interactions as well as a suggestion of a further Coriolis interaction with an unidentified state in CD_3H. Quack and co-workers identify a set of Fermi resonance interactions that explain the coarse structures observed in the vibrational overtone spectra of CF_3H[19] and $(CF_3)_3CH$[20] as well as CD_3H,[21] and they are able to reproduce the structure of the spectra using an effective Hamiltonian. Such models may be fairly general, and Hutchinson *et al.*[24] propose that similar non-linear resonances are generally important in the spectroscopy of alkanes.

The measurement and analysis of the vibrational-overtone spectrum of CF_3H[19] illustrate several aspects of the model of a zero-order state that is strongly coupled to a few background states that are, in turn, weakly coupled to a large number of other background states. Multiple broad features in the spectrum come from the interaction of the zero-order state with the strongly coupled Fermi resonant states. This interaction creates several resultant states that share the oscillator strength of the zero-order state. If these were all of the states in the molecule, each transition would be a narrow feature whose width would reflect the natural lifetime of the state. However, there are many weakly coupled background states whose interaction with the resultant states leads to the spectral width of the features. The strong interaction with the Fermi resonant states is built into the effective Hamiltonian of Quack and co-workers, but the weak coupling is only treated as producing much of the observed linewidth of the features. As both Dübal and Quack[19] and Sibert *et al.*[16] illustrate, knowing the interactions in the molecule allows the prediction of the time evolution of the zero-order state *if it were to be prepared*. The strong interactions control the short-time behaviour while the weaker interactions become important in the longer-time behaviour. An essential point in considering vibrational overtone transitions, which are usually labelled by the identity of their zero-order state, is that the optical excitation connects the unexcited state with the states, created by the interaction of the zero-order state with the other states in the molecule, that lie within the laser linewidth. The states that share most of the oscillator strength are those lying within the linewidth that are most

[24] J. S. Hutchinson, J. T. Hynes, and W. P. Reinhardt, *J. Chem. Phys.*, 1986, **90**, 3528.

strongly coupled to the zero-order state,[25] an example being those states that are in Fermi resonance with it.

The structure in the spectrum that comes from transitions from a single lower state to several different upper states is *homogeneous*.[22] Clearly, states produced by Fermi resonance interactions that couple the zero-order state to some of the background states are one source of homogeneous structure in vibrational overtone spectra. Another is the presence of very similar but inequivalent oscillators in the molecule.[26] The CH stretching vibrations of a methyl group in an alkane are illustrative. The three CH bonds in the methyl group are not exactly equivalent in many molecules because they are in slightly different environments at the equilibrium position of the methyl group. For example, one might lie in the plane of the two adjacent carbon–carbon bonds, and two might be located out of that plane.[26] The different interactions in these two environments lead to slightly different bond strengths and, hence, vibrational transition frequencies. Thus the vibrational overtone transitions occur at distinct wavelengths and appear as separate, or at least partially resolved, features in the spectrum. Thermal excitation populates higher-energy states of the methyl rotor and produces 'hot band' transitions that contribute to the observed width of a spectral feature, but the basic structure coming from the inequivalent hydrogens is, like all homogeneous structure, independent of temperature.

Spectral structure that arises from transitions originating in thermally populated initial states is *inhomogeneous*. The hot-band transitions of methyl rotor states described above, and all hot bands in general, are examples of inhomogeneous structure. In vibrational overtone spectra rotational contours are an example of inhomogeneous broadening, which in sufficiently high-resolution experiments produces structure on individual vibrational transitions. Because inhomogeneous structure is temperature dependent, measurements on cold samples can identify its contribution and, by removing it, uncover the homogeneous structure. One example is the dramatic change in the spectrum of the fifth vibrational overtone transition in CH_4 upon cooling the sample from room temperature to $77\,K$,[18] where the inhomogeneous contributions are substantially reduced.

A vibrational overtone spectrum of a room-temperature sample has both homogeneous and inhomogeneous contributions. The homogeneous structure is the more interesting because it mirrors the important couplings between the zero-order state and the background state and thus is a basis for understanding the intramolecular dynamics. However, the inhomogeneous structure is important for understanding the nature of the state prepared by vibrational overtone excitation. A hot-band transition may well prepare a molecule in a very different state than a nearby transition originating in the ground state. Studying vibrational photodissociation by means of vibrational

[25] J. M. Jasinski, J. K. Frisoli, and C. B. Moore, *Faraday Discuss. Chem. Soc.*, 1983, **75**, 289.
[26] J. S. Wong, R. A. MacPhail, C. B. Moore, and H. L. Strauss, *J. Phys. Chem.*, 1982, **86**, 1478; J. S. Wong and C. B. Moore, *J. Chem. Phys.*, 1982, 77, 603.

overtone excitation requires careful consideration of the consequences of both aspects of the spectra.

Vibrational Overtone Spectra

We obtain vibrational overtone spectra of both room-temperature and cooled molecules by monitoring the products of the vibrational overtone induced decomposition. The technique combines laser excitation of an overtone vibration with time-resolved spectroscopic product detection by either chemiluminescence or laser-induced fluorescence (LIF). The product yield as a function of the wavelength of the vibrational overtone excitation laser is the absorption spectrum of the molecules that decompose into chemiluminescent products, in the first detection scheme, or into the quantum state probed by the interrogating laser, in the second detection scheme. These approaches are applicable to molecules in low-pressure gases and in free-jet expansions, and we have studied both large molecules,[27-30] having over 50 vibrational degrees of freedom, and small molecules,[31-34] having only six vibrations, in both environments.

Spectra of Large Molecules.   A large molecule whose vibrational overtone spectroscopy we have studied extensively using chemiluminescence detection is tetramethyldioxetane (TMD).[27-29] It is a cyclic peroxide that decomposes over a barrier of approximately $110 \, kJ \, mol^{-1}$ to produce two acetone fragments:

$$\text{O—O} \longrightarrow \text{O} + \text{O}$$

As the energy-level diagram in Figure 1 shows, the energy barrier and the exothermicity are sufficient to form one of the fragments in an electronically excited state, a process that occurs with a quantum yield of 30% in solution.[35] We monitor the reaction by observing the emission from the electronically excited product following excitation of an overtone vibration.

The room-temperature vibrational overtone excitation spectrum of tetramethyldioxetane has a relatively complex structure that varies significantly with the excitation region. Figure 2 shows the total product chemiluminescence intensity for vibrational overtone excitation in the range 8300—14 300 cm^{-1}.[28] The three primary transitions, labelled $3\nu_{CH}$, $4\nu_{CH}$, and $5\nu_{CH}$, follow the simple Birge–Sponer relationship that is characteristic of local-mode spectra, $n\tilde{\nu}_{CH} = An + Bn^2$, where A–B is the mechanical frequency of

[27] B. D. Cannon and F. F. Crim, *J. Chem. Phys.*, 1981, **75**, 1752.
[28] E. S. McGinley and F. F. Crim, *J. Chem. Phys.*, 1986, **85**, 5741.
[29] E. S. McGinley and F. F. Crim, *J. Chem. Phys.*, 1986, **85**, 5748.
[30] T. R. Rizzo and F. F. Crim, *J. Chem. Phys.*, 1982, **76**, 2754.
[31] T. R. Rizzo, C. C. Hayden, and F. F. Crim, *J. Chem. Phys.*, 1984, **81**, 4501.
[32] T. M. Ticich, T. R. Rizzo, H.-R. Dübal, and F. F. Crim, *J. Chem. Phys.*, 1986, **84**, 1508.
[33] H.-R. Dübal and F. F. Crim, *J. Chem. Phys.*, 1985, **83**, 3863.
[34] L. J. Butler, T. M. Ticich, M. D. Likar, and F. F. Crim, *J. Chem. Phys.*, 1986, **85**, 2331.
[35] N. J. Turro, P. Lechtern, N. E. Schore, G. Schuster, H.-C. Steinmetzer, and A. Yekta, *Acc. Chem. Res.*, 1974, **7**, 97.

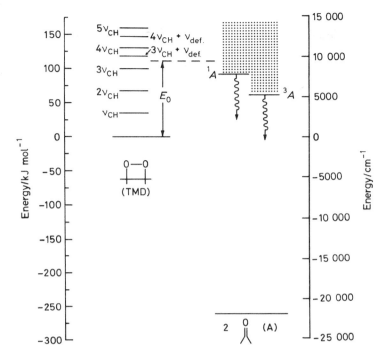

Figure 1 *Energy-level diagram for the vibrational overtone initiated decomposition of tetramethyldioxetane (TMD) to form electronically excited acetone (^1A and ^3A). The barrier to the reaction is E_0. The overtones of the CH stretching vibration are labelled nv_{CH}, and the combination bands with the methyl deformation are labelled $nv_{CH} + v_{def}$.*

(Reproduced with permission from *J. Chem. Phys.*, 1986, **85**, 5741)

the oscillator, B is the anharmonicity (defined as a negative quantity), and n is the local-mode vibrational quantum number. We assign the features to the second, third, and fourth CH stretching overtone transitions, respectively. Two weaker transitions, labelled $3v_{CH} + v_{def.}$ and $4v_{CH} + v_{def.}$, appear at approximately 1400 cm^{-1} higher energy than the $3v_{CH}$ and $4v_{CH}$ transitions. This separation is consistent with the features being combination transitions that excite either three or four quanta of the CH stretching vibrations and one quantum of a methyl deformation.[36,37]

The two prominent maxima in the spectra of the $5v_{CH}$ and the $4v_{CH}$ transitions are almost certainly examples of structure arising from inequivalent hydrogens in the methyl groups of TMD.[27,28,38] One test of this assignment is in the vibrational overtone excitation spectrum of TMD molecules cooled in a free-jet expansion.[28] Figure 3 shows the spectrum obtained in a pulsed expansion for excitation in the region of the $4v_{CH}$ transition. The

[36] R. Aroca and M. Menzinger, *Spectrosc. Lett.*, 1983, **16**, 945.
[37] H. L. Fang and R. L. Swofford, *J. Chem. Phys.*, 1980, **73**, 2607.
[38] E. S. McGinley, Ph.D. thesis, University of Wisconsin, 1985.

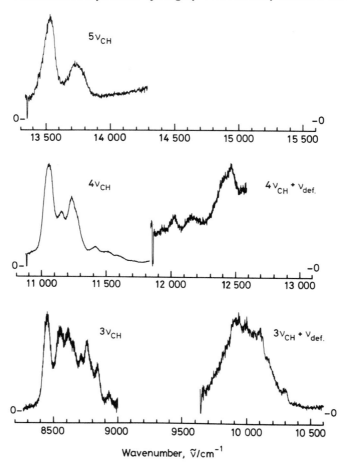

Figure 2 *Room-temperature vibrational overtone predissociation spectra of tetra-methyldioxetane. The signal is the chemiluminescence intensity normalized to the pulse energy of the excitation laser and has been adjusted to the same maximum intensity in each of the five wavelength regions shown. The zero of the signal, obtained by blocking the excitation laser, is shown for each spectrum. The spectra are taken at pressures of* 20 mTorr *or less*
(Reproduced with permission from *J. Chem. Phys.*, 1986, **85**, 5741)

expansion is modest, about 20 Torr* of helium through a 1 mm diameter nozzle, but it does cool the molecules enough to begin to differentiate the homogeneous and inhomogeneous components of the spectrum. Although relative intensities of some features change for the $4v_{CH}$ spectrum as inhomogeneous contributions are reduced upon cooling, *all of the features present in the room-temperature spectrum are present in the spectrum taken in*

---

* 1 Torr = 133.3 Pa

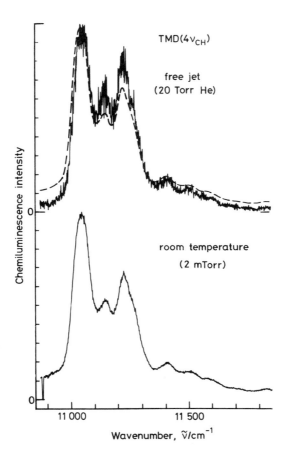

Figure 3 *Vibrational predissociation spectra of the third overtone of the CH stretching vibration of tetramethyldioxetane (4v_{CH}) in a free jet (upper trace) and at room temperature (lower trace). The broken line in the free-jet spectrum shows the room-temperature measurement for comparison. The carrier gas in the free-jet expansion is helium at a stagnation pressure of 20 Torr, and the total TMD pressure in the room-temperature measurement is 2 mTorr*
(Reproduced with permission from *J. Chem. Phys.*, 1986, **85**, 5741)

the free jet. This comparison shows that most of the structure is independent of temperature and, hence, homogeneous in origin.

The Fermi resonance interactions that have been implicated in other vibrational overtone spectra are the likely source of part of the homogeneous structure in the TMD spectrum. Because combination bands involving the methyl deformation are prominent in the vibrational overtone spectra of TMD and the methyl deformations occur at roughly half the frequency of the 2800 cm^{-1} CH stretching vibrations, we expect the two modes to be in Fermi resonance, in analogy to the situation with the CH bend and CH stretch in $(CF_3)_3CH$[20] and benzene.[16] Table 1 lists the

Table 1   *Calculated wavenumbers* (in cm^{-1}) *for the* CH *overtone transitions and for the corresponding lowest-order Fermi resonant transitions involving the methyl deformation*[a]

n	Transition		Separation[b]
	$n\tilde{v}_{CH}$	$(n-1)\tilde{v}_{CH} + 2\tilde{v}_{def.}$	δ
3	8451	8625	−174
4	11 047	11 331	−284
5	13 533	13 927	−394

[a] The transition wavenumbers are calculated using the Birge–Sponer parameters from ref. 27, A = 2983 cm^{-1}, B = −55.3 cm^{-1}, and a methyl deformation frequency of 1440 cm^{-1}. [b] $\delta = n\tilde{v}_{CH} - [(n-1)\tilde{v}_{CH} + 2\tilde{v}_{def.}]$

frequencies of the pure CH overtone transitions ($n v_{CH}$), calculated from the mechanical frequency and anharmonicity determined by a Birge–Sponer analysis of the CH stretching vibration in TMD,[27] and those of the corresponding lowest-order Fermi resonant combination bands [$(n-1)v_{CH} + 2v_{def.}$]. Although the Fermi resonance interaction shifts the transition frequencies,[19-22] their separation δ does serve as a rough measure of the extent of the interaction for comparable interaction matrix elements. As the values of δ in Table 1 show, the CH stretching state in the lowest-frequency region we observe, $3v_{CH}$, is separated from the Fermi resonant state having one quantum of stretching replaced by two quanta of methyl deformation by only 174 cm^{-1}, but the energy difference grows to almost 400 cm^{-1} for $5v_{CH}$. The resulting combination states should carry less of the CH stretching transition probability for the higher levels and, consequently, the structure should be less pronounced. This agrees with our observation that the structure that is prominent in the $3v_{CH}$ region has reduced intensity in the $4v_{CH}$ region and does not appear in the $5v_{CH}$ region (Figure 2). Thus, we are probably observing the consequences of Fermi resonance interactions between the CH stretching mode (the zero-order state) and the methyl deformation mode (the strongly coupled background state), which become less important for the higher CH vibrational levels because the larger energy separation leads to a weaker interaction.

The comparison of the room-temperature and free-jet spectra in the region of $5v_{CH}$ reveals additional changes. Figure 4 shows two $5v_{CH}$ spectra taken at room temperature and in a 100 Torr expansion of helium. Each of the features, which have relatively smooth envelopes at room temperature, partially separates into at least two components in the cooled sample. This argues that the features are *not* purely homogeneously broadened. Our measurement shows that *there is more than one component to each of the features* but does not determine their homogeneous widths. The observation of temperature-dependent features in the vibrational overtone spectrum of TMD in the region of the $5v_{CH}$ transition differs from the results of an earlier study by West *et al.*,[39] in which they concluded that the features do not

[39] G. A. West, R. P. Mariella, jun., J. A. Pete, W. B. Hammond, and D. F. Heller, *J. Chem. Phys.*, 1981, **75**, 2006.

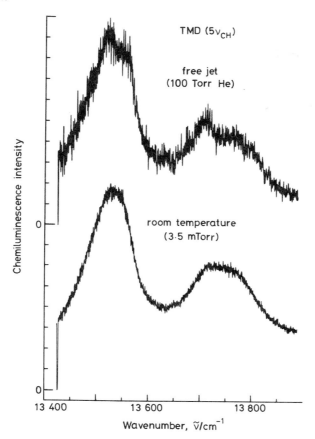

Figure 4 *Vibrational predissociation spectra of the fourth overtone of the CH stretching vibration of tetramethyldioxetane ($5\nu_{CH}$) in a free jet (upper trace) and at room temperature (lower trace). The carrier gas in the free-jet expansion is helium at a stagnation pressure of 100 Torr, and the total TMD pressure in the room-temperature measurement is 3.5 mTorr*
(Reproduced with permission from *J. Chem. Phys.*, 1986, **85**, 5741)

change with temperature and, thus, have a homogeneous width of 100 cm^{-1}. The data in Figure 4 indicate that the widths are no more than 50 cm^{-1} and are possibly less, since inhomogeneous contributions might remain. The structure that we observe in the free-jet spectrum for $5\nu_{CH}$ could have several origins. For example, it might come from combination bands that involve a quantum of a low-frequency motion, such as the ring deformation or methyl torsion, or arise from several inequivalent hydrogens that differ only slightly in their environments. Other resonant interactions involving the combination of two different vibrational modes, analogous to that postulated by Hutchinson for cyanoacetylene,[40] are possible as well.

[40] J. S. Hutchinson, *J. Chem. Phys.*, 1985, **82**, 22.

The preceding discussion shows that the vibrational overtone excitation spectra of tetramethyldioxetane exhibit features that are consistent with observations in several other large molecules. There is evidence for homogeneous structure from inequivalent methyl hydrogens and from Fermi resonance interactions between the zero-order CH stretching state and the state corresponding to the methyl deformation mode. In particular, the detuning of the resonance with increasing excitation explains much of the change in spectral structure in going from the region of the $3\nu_{CH}$ transition to that of the $5\nu_{CH}$ transition. The measurements on molecules cooled in a free-jet expansion show that there are inhomogeneous contributions to the room-temperature spectrum as well, and that a portion of the linewidth in the room-temperature $5\nu_{CH}$ spectrum arises from inhomogeneous contributions. As several studies demonstrate,[16-22] careful analysis of room-temperature measurements, particularly for molecules having isolated CH stretches, often sheds considerable light on the important couplings to the zero-order state. However, removing the inhomogeneous contributions by cooling the molecules is necessary to isolate the homogeneous structure. Eliminating the inhomogeneous components of the spectrum is no guarantee of observing isolated transitions to single eigenstates of the molecule. If, as can be the case in a large molecule, the number of background states leads to many states that have a significant oscillator strength lying within the bandwidth of the laser or if these states are separated by less than their natural linewidth, excitation of a single state is impossible. Because a larger number of states in a molecule makes such a situation more likely to occur, small molecules offer the best possibility for observing truly isolated vibrational overtone transitions and understanding them theoretically.

Spectra of Small Molecules. Hydrogen peroxide (HOOH) has proven to be an excellent molecule for highly resolved studies of vibrational overtone excitation and the resulting unimolecular reaction dynamics. Because it is a relatively simple molecule, *ab initio* calculations of portions of its potential-energy surface are feasible,[41,42] and its relatively high-frequency vibrations produce a small enough density of background states that high-resolution spectra can potentially isolate individual transitions. The one low-frequency vibration in the molecule, the torsion of the two OH groups about the O—O bond, complicates the vibrational overtone spectroscopy but, at the same time, makes it possible to extract the dependence of the torsional potential on the level of vibrational excitation.

The decomposition of hydrogen peroxide requires about 210 kJ mol^{-1} of energy and produces two readily detectable fragments:

$$HOOH \longrightarrow OH + OH$$

Figure 5 shows the energetics of the vibrational overtone initiated dissociation and the vibrational and rotational state-specific detection of the CH

[41] J. E. Carpenter and F. A. Weinhold, *J. Phys. Chem.*, 1986, **90**, 6405.
[42] L. B. Harding and A. F. Wagner, private communication.

fragment by LIF. Vibrational overtone excitation spectra come from varying the wavelength of the exciting laser while monitoring the LIF signal with the probe laser tuned to a particular rovibrational state of the OH fragment. Only the transition in the $6\nu_{OH}$ region has a photon energy that exceeds the strength of the OO bond, but two other processes permit decomposition following excitation with photons of lower energy.

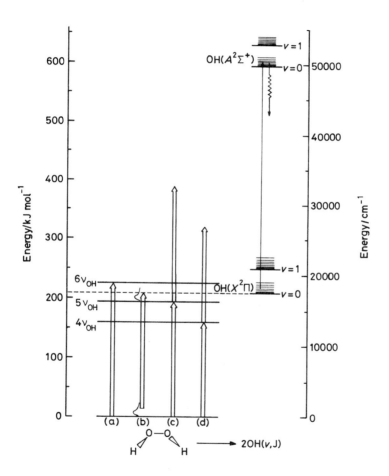

Figure 5 *Energy-level diagram for the one- and two-photon dissociation of hydrogen peroxide. The double arrow furthest left indicates the one-photon excitation of the fifth overtone vibration ($6\nu_{OH}$). The two middle double arrows show the two-photon excitation of a dissociative state through the fourth overtone vibration ($5\nu_{OH}$) as the intermediate state and the one-photon excitation of the fourth overtone vibration that leads to the 'thermally' assisted decomposition, respectively. The double arrow furthest right shows the two-photon excitation of a dissociative state through the third overtone vibration ($4\nu_{OH}$). The single arrows illustrate the laser-induced fluorescence detection of individual product vibrational rotational states*

One process is dissociation of thermally excited molecules. Although the photon energy for the $5\nu_{OH}$ transition is about 1100 cm^{-1} less than the dissociation energy, excitation of molecules that possess at least that much thermal energy leads to decomposition. The resulting excitation spectrum is biased in favour of hot-band transitions and excited initial rotational states and, in fact, can be analysed to extract information about the thermally excited OO stretching vibration.[32] The other process is a two-photon dissociation that uses the vibrational overtone transition as the first step.[43] For example, excitation of the $4\nu_{OH}$ transition adds far too little energy to the molecule for dissociation, but absorption of a second photon takes the molecule to a rapidly dissociating state from which it decomposes to produce OH fragments. Because the second transition is apparently unstructured, the excitation spectrum mirrors the first absorption step between the ground state and the bound vibrational overtone state. This two-photon dissociation process also becomes dominant in the $5\nu_{OH}$ region when high rotational states of the OH fragment are interrogated. Producing only one OH fragment in $N = 9$ requires about 1600 cm^{-1} of energy in addition to the 1100 cm^{-1} required, beyond the $5\nu_{OH}$ excitation, to produce the rotationless fragments (N is the total angular momentum less the electron spin). Because an insignificant number of molecules possess this much thermal energy, the two-photon dissociation pathway is responsible for all of the highly rotationally excited ($N > 7$) OH fragments produced by excitation in the $5\nu_{OH}$ region. The key to separating out this contribution is the detection of individual rotational states of the fragment. For both the $4\nu_{OH}$ and $5\nu_{OH}$ regions the first photon produces the vibrational overtone excitation spectrum and the second photon serves as part of the detection scheme by creating the easily detected OH fragment.

The vibrational overtone excitation spectra for the $4\nu_{OH}$, $5\nu_{OH}$, and $6\nu_{OH}$ regions shown in Figure 6 clearly exhibit extensive, but incompletely resolved, vibrational and rotational structure. The spectrum for $6\nu_{OH}$ is the one-photon vibrational dissociation spectrum and those for $5\nu_{OH}$ and $4\nu_{OH}$ are two-photon measurements, as described above. The coarse vibrational structure in all three spectra consists of a main band with an adjacent satellite and a third feature located from 250 to 400 cm^{-1} higher in energy, depending on the vibrational overtone excitation region. The separation of both the satellite and the high-energy feature from the main band increases for excitation of higher overtone vibrations. The wavenumbers of the main transitions follow the Birge–Sponer extrapolation from lower vibrational levels,[44] and we label them as the three overtone transitions, $4\nu_{OH}$, $5\nu_{OH}$, and $6\nu_{OH}$. Each feature has sharp but congested rotational structure that is most apparent in the $4\nu_{OH}$ region.

A possible origin of the satellite and high-energy features is Fermi resonance interactions of the type discussed above for the CH stretch and

[43] T. M. Ticich, H.-R. Dübal, M. D. Likar, L. J. Butler, and F. F. Crim, *J. Chem. Phys.*, to be submitted.

[44] T. R. Rizzo, C. C. Hayden, and F. F. Crim, *Faraday Discuss. Chem. Soc.*, 1983, **75**, 350.

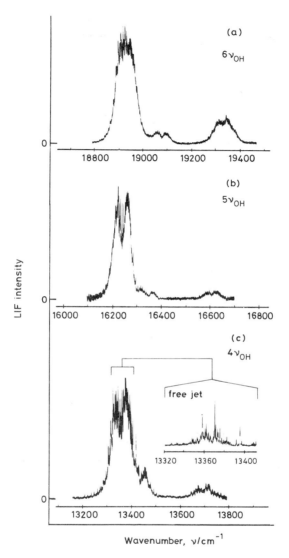

Figure 6 *Excitation spectra in the regions of the third (4v$_{OH}$), fourth (5v$_{OH}$), and fifth (6v$_{OH}$) overtone vibration transitions of hydrogen peroxide. All spectra come from monitoring a single rotational state of the OH decomposition product by laser-induced fluorescence while varying the wavelength of the vibrational overtone excitation laser. The spectra for the 4v$_{OH}$ and 5v$_{OH}$ regions are two-photon spectra, and that for the 6v$_{OH}$ region is a one-photon spectrum. The inset on the 4v$_{OH}$ spectrum shows the free-jet spectrum in the same region*

CH bends in alkanes. To test the possibility that such an interaction between the zero-order OH stretching state and an unidentified state produces the structure in the vibrational overtone excitation spectra of HOOH, we fit the structure in the $6\nu_{OH}$ region using this model and compare the predictions for $4\nu_{OH}$ and $5\nu_{OH}$ to the measurements. The Fermi resonance analysis consistently predicts red-shifted transitions that do not appear in the experimental spectra.[33] This result is also consistent with the analysis of the intramolecular dynamics of HOOH by Uzer *et al.*[45] in which they concluded that the Fermi resonances between the OH stretch and OOH bend are badly detuned in the spectral regions we have investigated.

We originally speculated that the high-energy feature was a combination band between the OH stretching state and the first excited torsional state of the molecule on the basis of the frequency shift from the main band.[31,44] The low frequency of the torsion ($\sim 200\,\mathrm{cm}^{-1}$) compared to that of the stretching vibration ($\sim 3600\,\mathrm{cm}^{-1}$) suggests developing a detailed spectroscopic model by adiabatically separating the co-ordinates for these two vibrations.[33] This approach, which is formally similar to the familiar Born–Oppenheimer separation of nuclear and electronic motion,[46] has been applied to a variety of molecules having disparate vibrational frequencies.[47] The adiabatic wavefunctions are:

$$\Psi_{nm}(s,\chi) = \psi_{nm}(\chi)\varphi_n(s,\chi) \tag{1}$$

where n and m are the quantum numbers for the OH stretching vibration and the torsion, respectively, and s and χ are the corresponding co-ordinates. The $\varphi_n(s,\chi)$ are the eigenfunctions of the Hamiltonian operator for the stretching vibration at a fixed torsional angle, and the corresponding eigenvalues at different torsional angles constitute the effective torsional potential. This effective torsional potential, $V_n(\chi)$, which is parametric in the vibrational quantum number n, describes the torsional motion for a particular level of OH stretching excitation within the limitations of the adiabatic separation of the two motions. The lowest curve in Figure 7 shows the torsional potential for the vibrational ground state extracted from far-infrared spectral data.[48,49] There is a relatively small barrier, $V_0^{trans} = 386\,\mathrm{cm}^{-1}$, at the *trans* position of the OH groups ($\chi = \pi$) and a much larger one, $V_0^{cis} = 2460\,\mathrm{cm}^{-1}$, at the *cis* position of the OH groups ($\chi = 0$). The *cis* barrier produces a very small splitting between the torsional levels, and the spectroscopy is only sensitive to the magnitude of the *trans* barrier.

The absorption cross-section for a transition between an initial vibrational–torsional state $|n''m''>$ and a final state $|n'm'>$ is proportional to the square of the transition moment, $<n'm'|\hat{\mu}|n''m''>$, and, for a dipole moment operator that is separable in the stretching and torsional co-ordinates, the

[45] T. Uzer, J. T. Hynes, and W. P. Reinhardt, *Chem. Phys. Lett.*, 1985, **117**, 600; *J. Chem. Phys.*, **85**, 5791.

[46] M. Born and K. Huang, 'Dynamical Theory of Crystal Lattices', Clarendon, Oxford, 1954.

[47] P. R. Bunker, *J. Mol. Spectrosc.*, 1980, **80**, 422; H.-R. Dübal and M. Quack, *Chem. Phys. Lett.*, 1982, **90**, 370; I. M. Mills, *J. Phys. Chem.*, 1984, **88**, 532.

[48] R. H. Hunt, R. A. Leacock, C. W. Peters, and K. T. Hecht, *J. Chem. Phys.*, 1965, **42**, 1931.

[49] R. L. Redington, W. B. Olson, and P. C. Cross, *J. Chem. Phys.*, 1962, **36**, 1311.

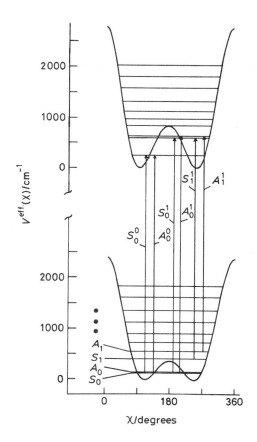

Figure 7 *Effective torsional potential for hydrogen peroxide extracted from the analysis of the vibrational overtone spectrum using an adiabatic separation of the OH stretching and torsional vibrations. The vertical arrows indicate some allowed vibrational–torsional transitions for hydrogen peroxide and are labelled by the symmetry of the torsional wavefunction with respect to the trans barrier and the quantum numbers of the upper and lower torsional states*
(Reproduced with permission from *J. Chem. Phys.*, 1985, **83**, 3863)

cross-section is proportional to the overlap of the torsional wavefunctions in the two vibrational states:

$$G \propto |<n'|\hat{\mu}s|n''>|^2| <m'|m''>|^2 \qquad (2)$$

The overlap integral of the torsional wavefunctions is analogous to the Franck–Condon factor in electronic spectroscopy and strongly influences the intensities of the transitions between particular vibrational levels. For example, if the torsional potentials in the two vibrational states are identical, the orthogonality of the torsional wavefunctions makes the overlap integral vanish unless the transition connects the same torsional states, $m' = m''$. The overlap integral imposes symmetry restrictions on the allowed transitions as well. Each torsional level, labelled S or A in Figure 7, is either symmetric or antisymmetric with respect to the *trans* barrier, and, because of the overlap integral, only transitions that connect levels of the same symmetry have non-zero cross-sections.

The magnitude of the *trans* barriers in the torsional potentials for the vibrationally excited and unexcited molecules determines the energy levels and, hence, the frequencies of the transitions. Because the far-infrared spectroscopic data establish the potential for the unexcited molecule,[48,49] it is possible to extract the parameters for the torsional potential of the vibrationally excited molecule by fitting the vibrational overtone spectra. Our approach is to solve the Schrödinger equation for the torsional energy levels numerically and to adjust the parameters in the potential, the *trans* barrier height and the equilibrium angle, to obtain the best fit to the observed spectra.[33]

The upper potential in Figure 7 is the result of fitting the data for the $6\nu_{OH}$ regions of HOOH. The transitions that produce the three features in the spectra are indicated in the figure. The main feature comprises transitions (S_0^0 and A_0^0) between the lowest torsional levels in the ground and vibrationally excited states. The satellite comes from hot-band transitions (S_1^1 and A_1^1) between the first excited torsional states in the vibrationally unexcited molecule and the corresponding states in the vibrationally excited molecule. The higher-energy feature is the combination band (S_0^1 and A_0^1) from the initially unexcited state to one in which the stretching vibration and one quantum of torsion are excited. Using this model, we extract the energy difference between the first torsional levels of a particular symmetry and, thus, the height of the *trans* barrier from the vibrational overtone spectra. As Figure 7 shows, the main band and the high-energy feature are separated by the energy difference between the torsional levels in the vibrationally excited molecule, and simple inspection of the spectra reveals its variation with excitation level.

The series of vibrational overtone spectra for the $4\nu_{OH}$, $5\nu_{OH}$, and $6\nu_{OH}$ regions (Figure 6) show the qualitative change in the height of the *trans* barrier in the adiabatic torsional potential. Both the high-energy feature and the satellite move away from the main feature in going from $4\nu_{OH}$ to $6\nu_{OH}$. Thus, the *trans* barrier must be increasing in the vibrationally excited molecule. Fitting the spectra quantitatively shows that the *trans* barrier increases by about $75\,\text{cm}^{-1}$ for each OH stretching vibrational quantum. The increase in the barrier height with increasing vibrational excitation comes from the change in the interaction of the orbitals on the oxygen atoms as the extension of the OH bond increases. *Ab initio* calculations of the energies of the *trans* and equilibrium configurations by Carpenter and Weinhold[41] reproduce the magnitude of this effect when the energies are averaged over the vibrational motion of the OH bond. They interpret the increasing barrier height as arising from a stabilization of the equilibrium configuration by improved overlap of the antibonding orbital from one OH bond with the lone-pair orbital on the other oxygen.[41,50]

Incompletely resolved rotational structure is apparent in all of the vibrational overtone excitation spectra in Figure 6. The abundance of transitions arising from thermally populated states prevents observation of single rotational transitions or detailed analysis of the rotational structure in the spectrum.[33] Reducing the number of initially populated states by cooling the molecules in a supersonic expansion[51] drastically alters this situation. Figure 8a shows the low-resolution ($\Delta\tilde{v} \simeq 0.3 \, cm^{-1}$) room-temperature two-photon vibrational overtone excitation spectrum in the $4v_{OH}$ region along with that obtained in a pulsed free-jet expansion of 300 Torr of helium through a 1 mm diameter nozzle.[34] The cooling in the expansion produces a great simplification of the spectrum. The width of the rotational contour diminishes to about $30 \, cm^{-1}$ from $100 \, cm^{-1}$, and the highly congested spectrum is reduced to about a dozen features. The variation in the intensities of the two prominent features at 13 368 and 13 357 cm^{-1} with expansion conditions suggests that they originate in the lowest symmetric and antisymmetric torsional states, which are separated by only $11.4 \, cm^{-1}$ in the ground vibrational state.[33] The high-resolution ($\Delta\tilde{v} = 0.05 \, cm^{-1}$) spectra (Figure 8c) show that each of the features is well separated and has a width of between 0.08 and $0.13 \, cm^{-1}$. Careful investigation of the changes in the shapes of these features with the extent of cooling in the expansion indicates that they contain some residual inhomogeneous structure. This may well be from very closely spaced Q-branch transitions that originate in the few lowest rotational states of HOOH.

These vibrational overtone excitation spectra of cold molecules provide an opportunity to analyse the spectra in enough detail to discover the couplings between the zero-order state and the backgound states. Hydrogen peroxide has such a low density of vibrational states in the region of the $4v_{OH}$ excitation, about 4 per cm^{-1},[34] that the presence of a large number of couplings seems unlikely. The most interesting question concerns the origin of the sharp features in the excitation spectrum. Many certainly come from the torsional and rotational states that remain populated in the supersonic expansion, but some could result from the interaction of the zero-order state with background states. Detailed spectroscopic analyses of these spectra can test this possibility. An example of such behaviour occurs in stimulated emission-pumping experiments on formaldehyde.[52] Sharp features that are not part of the zero-order description appear in the spectrum because Coriolis and perhaps other unidentified interactions couple states in the molecule. The virtue of high-resolution studies on small molecules is that the intramolecular interactions do not lead to a broad homogeneous linewidth

[50] T. K. Brunck and F. A. Weinhold, *J. Am. Chem. Soc.*, 1979, **101**, 1700.
[51] T. E. Gough, R. E. Miller, and G. Scoles, *Appl. Phys. Lett.*, 1977, **30**, 338; *J. Mol. Spectrosc.*, 1978, **72**, 124; R. E. Smalley, L. Wharton, and D. H. Levy, *Acc. Chem. Res.*, 1977, **10**, 139.
[52] E. Abramson, R. W. Field, D. Imre, K. K. Innes, and J. L. Kinsey, *J. Chem. Phys.*, 1984, **80**, 2298.

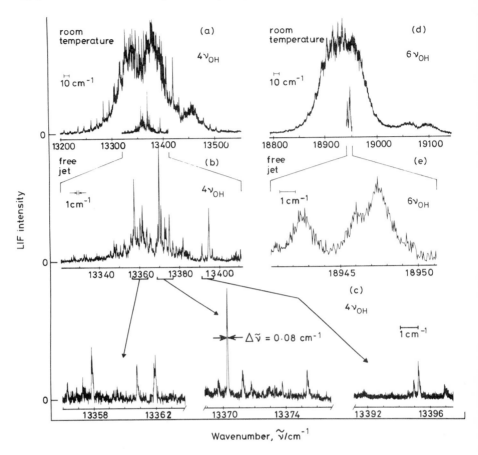

Figure 8 *Spectra of vibrational overtone transitions to bound ($4\nu_{OH}$) and dissociative ($6\nu_{OH}$) states of hydrogen peroxide in room-temperature samples and in a supersonic expansion. The top spectrum in (a) is the spectrum in the $4\nu_{OH}$ region measured in a room-temperature sample, and that below it is the one measured in a supersonic expansion. An expanded spectrum is shown in (b), and a high-resolution ($\Delta\tilde{\nu} \simeq 0.05$ cm^{-1}) one is shown in (c). The top spectrum in (d) is the spectrum measured in the $6\nu_{OH}$ region in a room-temperature sample, and that below it is the one measured in a supersonic expansion. The expanded high-resolution spectrum in (e) ($\Delta\tilde{\nu} = 0.045$ cm^{-1}) shows that the features are considerably broader than in the bound-state spectrum (c)*

(Reproduced with permission from *J. Chem. Phys.*, 1986, **85**, 2331)

but rather, because of the small number of states involved, produce discrete homogeneous structure whose location can potentially be analysed to identify the couplings.

A qualitative manifestation of the consequences of changing the number of states coupled to the zero-order state comes from exciting above the dissociation limit, where the continuum of unbound states becomes available. The free-jet spectrum in the region of the $6v_{OH}$ transition is illustrative. Figure 8d shows that cooling HOOH in a free-jet expansion greatly simplifies the room-temperature spectrum in the $6v_{OH}$ region, but the high-resolution spectrum (Figure 8e) shows that the features are substantially broader than for excitation in the $4v_{OH}$ region. Because the density of bound states has only increased by a factor of about three from the energy of $4v_{OH}$ to that of $6v_{OH}$,[34] it is unlikely that most of the width of the apparently isolated feature near 18 942.5 cm^{-1} comes from interaction of the zero-order state with bound states. Rather, it is much more probable that the 1.5 ± 0.3 cm^{-1} linewidth primarily reflects the coupling into the continuum and, hence, the dissociation lifetime of the highly vibrationally excited molecule.

The studies of vibrational overtone spectroscopy discussed in this section bear most closely on the question of what is excited to initiate the vibrational photodissociation process. Both homogeneous and inhomogeneous structure are important in vibrational overtone spectra of room-temperature samples, as the measurements on tetramethyldioxetane and hydrogen peroxide demonstrate. Cooling the molecules in a free-jet expansion can reduce or even eliminate the complication of inhomogeneous structure. The two types of homogeneous structure that seem most important in our example of a large molecule, TMD, come from the presence of inequivalent, but very similar, CH oscillators in the methyl groups and from Fermi resonance interactions that couple the zero-order CH stretching state with methyl deformation vibrations. Much more vibrational and rotational structure is apparent in the room-temperature spectrum of our prototypical small molecule, HOOH. A simple model which adiabatically separates the low-frequency torsional vibration and the high-frequency OH stretching vibration explains the coarse vibrational structure in the spectrum and its variation with excitation energy. The most important consequence of this analysis for vibrational dissociation dynamics is the clear identification of spectral regions in which a quantum of torsional excitation is created along with the OH stretching excitation. Cooling hydrogen peroxide in a free-jet expansion removes a great deal of inhomogeneous structure and, consequently, reveals a simplified vibrational overtone excitation spectrum. Such data offer an opportunity for detailed spectral analysis to identify the operative couplings within the molecule and provide a means of highly selective excitation in which the total angular momentum is quite small and the initial vibrational excitation energy is well defined. These spectral measurements on large and small molecules set the stage for probing the dissociation dynamics by measuring dissociation rates and product quantum-state distributions.

3   Dissociation Rates

Measurements of the dissociation rates of highly vibrationally excited molecules prepared by vibrational overtone excitation offer an excellent opportunity for comparison with the detailed predictions of theory.[53] Because the decomposition of isolated energized molecules is a fundamental chemical event, it has received extensive theoretical[54] and experimental[55] attention for many decades. Detailed statistical theories[54,56,57] predict the rate constant $k(E,J)$ as a function of the total energy E and total angular momentum J, and comparison with theory demands equally detailed measurements. Vibrational overtone excitation adds a known energy increment to the molecule and, in a jet-cooled sample where the range of thermal energies is small, prepares the molecule with a precisely known total energy. Highly resolved vibrational overtone excitation of a jet-cooled sample, in which the molecules have little rotational angular momentum and individual rotational transitions can be excited, is a means of preparing molecules with precisely known values of E and J. Vibrational overtone excitation even offers the possibility of mode-selective excitation in molecules where the intramolecular couplings are favourable.

We have taken two approaches to determining the unimolecular reaction rates of molecules with excited overtone vibrations. One is quite direct in that we spectroscopically monitor the appearance of reaction products as a function of time. Because the 10 ns duration of our laser pulses determines the time resolution of this method, we apply it to large molecules, having about 50 vibrational degrees of freedom, whose unimolecular reaction rates are slow enough to be resolved.[27,29,30] Laser pulses of this duration are not a fundamental limitation to the technique of directly observing the time evolution of dissociation products, and Scherer *et al.*[58] have recently demonstrated its application to a small molecule using picosecond techniques on a room-temperature sample. Our second approach is a more inferential scheme in which we attempt to observe single, homogeneous lines in a vibrational overtone spectrum and determine the unimolecular reaction rate from the widths of these features.[34] This technique is complementary to the first since it is best applied to small molecules that react so rapidly that the resulting linewidth substantially exceeds our limiting laser resolution of about 0.05 cm^{-1}. Performing both experiments on molecules cooled in a free-jet expansion is an important aspect of our approach since it reduces the distribution of initial thermal energies in large molecules[28] and provides the spectral simplification required to observe isolated features in small molecules.[34]

[53] E. S. McGinley, L. J. Butler, T. M. Ticich, M. D. Likar, and F. F. Crim, *J. Chim. Phys.*, in press.
[54] P. J. Robinson and K. A. Holbrook, 'Unimolecular Reactions', Wiley, New York, 1972.
[55] F. F. Crim, *Annu. Rev. Phys. Chem.*, 1984, **35**, 657.
[56] M. Quack and J. Troe, *Ber. Bunsenges. Phys. Chem.*, 1972, **78**, 240.
[57] D. M. Wardlaw and R. A. Marcus, *Chem. Phys. Lett.*, 1984, **110**, 230; *J. Phys. Chem.*, 1985, **83**, 3462.
[58] N. F. Scherer, F. E. Doany, A. H. Zewail, and J. W. Perry, *J. Chem. Phys.*, 1986, **84**, 1932.

Direct Measurements

The energy-level diagram in Figure 1 illustrates the application of this technique to tetramethyldioxetane. Excitation in the regions of the two overtone transitions $4v_{CH}$ and $5v_{CH}$ and of the combination band $4v_{CH} + v_{def.}$ adds sufficient energy to the molecule for it to decompose to produce electronically excited acetone. (The reaction of molecules possessing sufficient thermal excitation to decompose following excitation in the $3v_{CH}$ region and the reaction of molecules excited in the lower of the two combination bands $3v_{CH} + v_{def.}$ are so slow that collisional quenching dominates the time evolution even at pressures of only 10 mTorr.) Observing the time evolution of the emission from the chemiluminescent product is a means of directly monitoring the progress of the vibrational overtone initiated dissociation. The total energy of a reacting molecule is the energy added by the photon plus any thermal energy that the molecule possessed prior to excitation. We know the energy increment, the photon energy hv, but are ignorant of the amount of thermal excitation that a particular molecule possesses. Thus, the distribution of initial energies limits our knowledge of the internal energy of the ensemble of reacting molecules. Overcoming this limitation by reducing the range of initial energies is one of the primary advantages of conducting vibrational overtone initiated dissociation measurements on molecules cooled in a supersonic expansion.

Figure 9 shows the time evolution of the product emission in a low-pressure room-temperature sample of TMD in which 10 ns laser pulses have excited the most prominent features in the $4v_{CH}$ and $5v_{CH}$ regions and in the combination band $4v_{CH} + v_{def.}$ region. The growth of the emission carries the information on the unimolecular reaction rate of the vibrationally excited TMD molecules, and the subsequent decay of the chemiluminescence reflects the photophysics and collisional quenching of the electronically excited acetone product. Haas and co-workers have studied the excited acetone produced both by infrared multi-photon-initiated reaction of TMD[59] and by ultraviolet excitation of acetone.[60] Their work and that of Copeland and Crosley[61] show that the complex double-exponential decay arises from the mixture of singlet and triplet character of acetone molecules formed above the origin of the singlet state.

The product time evolution in these experiments reflects the unimolecular reaction rate of a selectively excited molecule. However, the averaging of the energy-dependent reaction rate constant $k(E)$ over the distribution of initial thermal vibrational energies in the room-temperature sample complicates the extraction of a unimolecular reaction rate constant from the data. The products do not appear with the single-exponential growth that would arise from reaction at one energy but rather follow a more complex time evolution because of the distribution of initial energies in the reacting molecules. The

[59] S. Ruhman, O. Anner, and Y. Haas, *J. Phys. Chem.*, 1984, **88**, 5162.
[60] G. D. Greenblatt, S. Ruhman, and Y. Haas, *Chem. Phys. Lett.*, 1984, **112**, 200; O. Anner, H. Auckermann, and Y. Haas, *J. Phys. Chem.*, 1985, **89**, 1336.
[61] R. A. Copeland and D. R. Crosley, *Chem. Phys. Lett.*, 1985, **115**, 362.

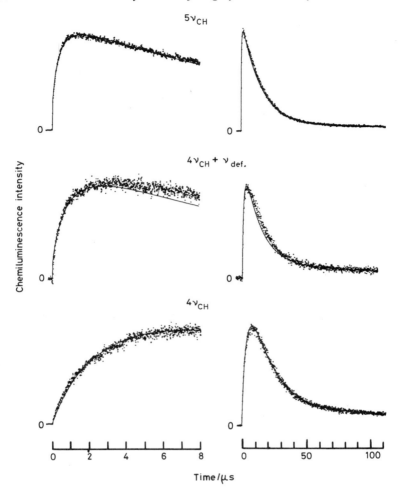

Figure 9 *Temporal evolution of the product chemiluminescence following excitation of the $4\nu_{CH}$, $4\nu_{CH} + \nu_{def.}$, and $5\nu_{CH}$ vibrational overtone transitions in tetramethyldioxetane. The data are at a total pressure of 10 mTorr and are shown for two time ranges spanning 8.0 μs and 110 μs, respectively. The solid lines are the results of the model calculations described in the text*
(Reproduced with permission from *J. Chem. Phys.*, 1986, **85**, 5748)

actual growth of the signal is the average of the simple exponential behaviour for a molecule that initially possesses thermal vibrational energy ε, to which the photon energy $h\nu$ is added to produce a total energy of $E = h\nu + \varepsilon$, over the distribution of thermal vibrational energy $P(\varepsilon)$:[27,29,38]

$$\int_0^\infty \exp[-k(h\nu + \varepsilon)t]P(\varepsilon)d\varepsilon \tag{3}$$

We have previously calculated $k(E)$ using RRKM theory and averaged the corresponding time evolution over a thermal distribution for comparison to room-temperature data.[27] However, reducing the range of initial energies by cooling the reactants in a free-jet expansion provides a more direct comparison. Our approach[29] is to reduce the thermal energy to the point that we can obtain a direct measure of the reaction rate for an ensemble of molecules at energies that are very near to the photon energy and then to use the resulting rate constant measurement to 'calibrate' our RRKM calculation. We can average this calculation, which is adjusted to agree with the measurement on a cooled sample, for comparison to room-temperature data at the energies of other vibrational overtone excitations.

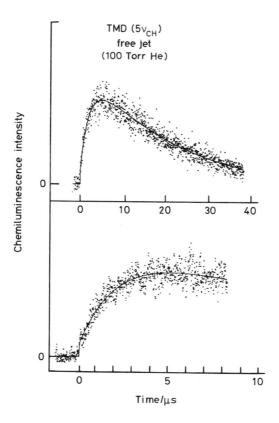

Figure 10 *Temporal evolution of the product chemiluminescence following excitation of the $5\nu_{CH}$ vibrational overtone transition in a free-jet expansion of approximately 0.2 Torr of tetramethyldioxetane in 100 Torr of helium. The upper portion of the figure displays only every fourth data point. The solid line is the non-linear least-squares fit of the sum of two exponentials to the data and has rate constants $k_{uni.} = 0.52\ \mu s^{-1}$ and $k_2 = 0.051\ \mu s^{-1}$*
(Reproduced with permission from *J. Chem. Phys.*, 1986, **85**, 5748)

The crucial measurement is the time evolution of the chemiluminescent product from the unimolecular decomposition of TMD initiated by excitation of the fourth CH stretching overtone ($5v_{CH}$) transition in a sample cooled in a pulsed free-jet expansion of 100 Torr of helium, as shown in Figure 10.[29] The relatively modest expansion through a 1 mm diameter nozzle does not collapse the distribution of initial internal energies $P(\varepsilon)$ to the delta function of an ideal experiment, but it does substantially reduce the range of energies over which the time evolution must be averaged. Thus, we approximate the average in equation (3) as a measurement at exactly the photon energy and fit the data as a single exponential rise of the form $\exp[-k(hv)t]$, as shown by the solid line in the figure. This gives a rate constant for unimolecular reaction of TMD initiated by excitation of the $5v_{CH}$ transition of $k(5v_{CH}) = (0.52 \pm 0.10)\,\mu s^{-1}$. This measurement serves as the calibration point for an RRKM calculation of the energy-dependent unimolecular reaction rate constant $k(E)$ for TMD.[29] The calculation is surprisingly robust. Because the shape of the $k(E)$ curve is very nearly the same for reasonable values of the threshold energy and low-frequency vibrations in the critical configuration and energized molecule,[29,38] the single measurement at $5v_{CH}$ constrains the calculation rather closely. This

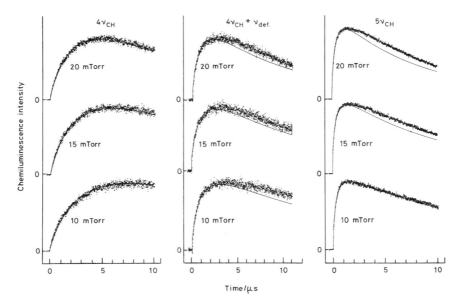

Figure 11 *The temporal evolution of the product chemiluminescence following excitation of the $4v_{CH}$, $4v_{CH} + v_{def.}$, and $5v_{CH}$ vibrational overtone transitions in a room-temperature sample of tetramethyldioxetane at three different pressures. The solid lines are the results of the model that incorporates the rate constant calculated using RRKM theory*

(Reproduced with permission from *J. Chem. Phys.*, 1986, **85**, 5741)

'calibrated' RRKM calculation is the basis for comparison of our more extensive data on room-temperature samples with theory.

Averaging the RRKM calculation over the thermal distribution of energies [equation (3)] predicts the time evolution of the product chemiluminescence. Figure 11 compares the averaged calculation (solid line) with the room-temperature measurements on three different vibrational overtone transitions. (The simulation also includes the small effect of collisional quenching of the vibrationally excited molecules prior to reaction.[29]) *The averaged RRKM calculation reproduces the growth of the signal quite well.* Slight differences between the calculation and measurements appear at longer times, when the photophysics of the electronically excited acetone product controls the time evolution, because our approximate treatment of the behaviour of the excited product is incomplete.[29,38] The growth of the signal reflects the unimolecular reaction, and the comparison with the calculation in this region shows that the data are entirely consistent with a statistical (RRKM) model of the dissociation.

Indirect Measurements

The spectrum of the $6\nu_{OH}$ vibrational overtone region for hydrogen peroxide molecules cooled in a supersonic expansion, shown in Figure 8e, is the means by which we infer the dissociation rate of the excited molecule. The width of the features in the region of the $6\nu_{OH}$ transition, which excites the HOOH above its dissociation threshold, is significantly greater than that of features in the region of the $4\nu_{OH}$ transition, which adds too little energy to dissociate the molecule. If, as discussed above, the $1.5 \pm 0.3\,\text{cm}^{-1}$ width of the feature at $18\,942.5\,\text{cm}^{-1}$ arises entirely from coupling into the continuum, the corresponding dissociation lifetime is $\tau = 3.5\,\text{ps}$. Because of the possibility of homogeneous coupling to bound states or a small amount of residual inhomogeneous structure contributing to the width, this is strictly only a lower limit to the lifetime, and, hence, the upper limit on the dissociation rate constant for $HOOH(6\nu_{OH})$ is $k(6\nu_{OH}) = 2.9 \times 10^{11}\,\text{s}^{-1}$. The cooling in the supersonic expansion ensures not only that this is the rate constant for HOOH molecules that have a total energy equal to the photon energy but also that it is for the decomposition of molecules possessing very little angular momentum. This closely approaches the desired measurement of the detailed rate constant $k(E,J)$ for comparison to the predictions of statistical theories.

The rate constant inferred from the linewidth is consistent with two rather different calculations of the lifetime of $HOOH(6\nu_{OH})$. It agrees with both a statistical calculation,[32,62] which assumes complete energy redistribution, and a trajectory calculation,[45] which finds that energy randomization is

[62] L. D. Brouwer, C. J. Cobos, J. Troe, H.-R. Dübal, and F. F. Crim, *J. Chem. Phys.*, 1987, **86**, in press.

incomplete. The applicability of a statistical theory to such a small molecule is by no means obvious, and a time-resolved measurement of the unimolecular reaction rate constant for HOOH($6\nu_{OH}$) would provide a particularly useful comparison to the one that we inferred from the linewidth measurement. Determining the limits of applicability of statistical models in hydrogen peroxide requires other diagnostics of the dynamics of the decomposition process, and the product-state distributions for HOOH cooled in a free-jet expansion are likely to be helpful in this regard.[63]

4   Product-State Distributions

The distribution of the products of a photodissociation among their quantum states is potentially a sensitive probe of the decomposition dynamics and is an important point of comparison with theoretical predictions. Our measurements of the populations of the rotational states of the OH product of the vibrational overtone initiated decomposition of hydrogen peroxide both in room-temperature gases and in free-jet expansions are a means of distinguishing different statistical theories of unimolecular decomposition. They may also provide data that explore the limits of statistical descriptions in such a small molecule. Product-state distributions are useful for qualitative comparisons between photodissociation processes as well. For example, the extent of vibrational excitation of the OH product of the two-photon dissociation of HOOH through a vibrational overtone state signals a significant difference between the dissociation dynamics for the vibrationally mediated case and for one-photon excitation directly to the electronically excited state.

Excitation of the fifth vibrational overtone transition ($6\nu_{OH}$) in HOOH prepares the molecule with about 20 kJ mol^{-1} of energy in excess of that required to break the O—O bond, as the energy-level diagram of Figure 5 illustrates. However, excitation of the fourth vibrational overtone transition adds insufficient energy for reaction, and only those molecules that possess some thermal vibrational or rotational energy are able to dissociate after absorption of a single photon in the region of the $5\nu_{OH}$ transition. Because the populations of the thermally excited states decrease rapidly with increasing energy, most of the reacting molecules are prepared very near the threshold energy for the dissociation in this case. Consequently, this thermally assisted reaction preferentially probes the dissociation at threshold, where the predicted product-state distributions are quite sensitive to the details of the theory. In the measurements, we tune the excitation laser to a particular feature in the vibrational overtone excitation spectrum and vary the wavelength of the probe laser to obtain the LIF excitation spectrum of the OH fragments. Converting these spectra to relative quantum-state populations gives the product-state distributions.[31]

The distributions of products among their rotational states for excitation

---

[63]   T. M. Ticich, L. J. Butler, M. D. Likar, and F. F. Crim, to be published.

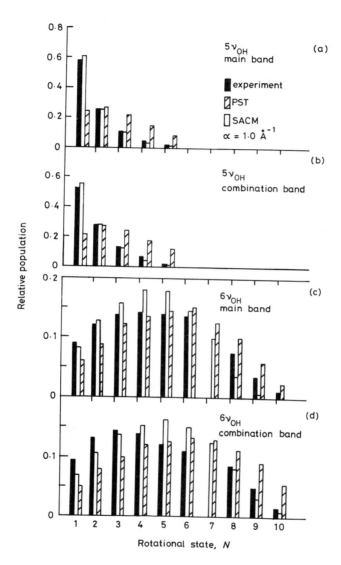

Figure 12 *Comparison of the observed product-state distributions (solid bars) and those calculated using phase space theory (hatched bars) and the statistical adiabatic channel model (open bars) for excitation of the main (a) and combination (b) bands in the fourth overtone ($5\nu_{OH}$) region and for excitation of the main (c) and combination (d) bands in the fifth overtone ($6\nu_{OH}$) region*

(Reproduced with permission from *J. Chem. Phys.*, 1986, **84**, 1508)

near threshold ($5v_{OH}$) and well above threshold ($6v_{OH}$) differ substantially. Figure 12 collects the experimental results (solid bars) for excitation of both the main bands and torsional combination bands in the region of the fourth and fifth vibrational overtone transitions.[32] The lowest rotational state is the most highly populated for fragments produced by excitation in the region of the $5v_{OH}$ transitions. Roughly half of the product molecules are born in $N = 1$ following excitation of either the main band or the combination band, and the population decreases monotonically with increasing rotational quantum number. The distribution for the combination band is shifted very slightly toward larger quantum numbers compared to that for the main band because of the $380 \, cm^{-1}$ of additional excitation energy provided by the more energetic photon.

The most populated rotational state in the fragments produced by excitation in the region of the $6v_{OH}$ transition is about $N = 4$ or 5, and the population is distributed over many more states than for the products resulting from excitation of the $5v_{OH}$ transition. Less than 20% of the products are formed in any one rotational state, and population appears in all the accessible states up to the energetic limit of the excitation photon. There is a subtle, but experimentally reproducible, difference between the product-state distributions resulting from excitation in the main band (Figure 12c) and in the combination band (Figure 12d) in the region of the $6v_{OH}$ transition. On a statistical basis, the larger energy of the excitation photon should shift the product-state distribution to slightly higher quantum numbers, but the experimental results show a change in the shape of the distribution in which the populations in the lower and higher quantum states increase while those of the intermediate states decrease compared to the distribution obtained for excitation in the main band. Exciting the combination band not only adds more energy but prepares a different zero-order nuclear motion as well. The observed differences in the product-state distributions may reflect the special role of the torsion in the dissociation because of a dynamical restriction that prevents it being completely coupled with the other degrees of freedom during the decomposition.

The measured product-state distributions at different levels of excitation permit a detailed comparison with the predictions of two statistical theories of unimolecular decomposition, the phase space theory[64] (PST) and the statistical adiabatic channel model[56] (SACM). Both models explicitly conserve energy and angular momentum but differ in their treatment of the interaction potential between the separating fragments. In the PST, bending vibrations in the molecule that are destined to become rotations in the fragments become completely free rotors as soon as the fragments begin to separate. This is the essence of the 'loose transition state' of the PST. In the SACM, the gradual reduction of the bending force constant during the separation of the fragments enters explicitly through a single additional

---

[64] P. Pechukas and J. C. Light, *J. Chem. Phys.*, 1965, **42**, 3281; P. Pechukas, J. C. Light, and C. Rankin, *J. Chem. Phys.*, 1966, **44**, 794.

parameter α. The consequence of the bending force constant changing in the course of the separation is that 'adiabatic barriers' appear along the separation co-ordinate for some initial states, particularly those that correlate with products having a large amount of rotational excitation, and close the decomposition channel.[31,56] As a result, the SACM predicts less rotational excitation in the products than does the PST.

The parameters required to calculate the probability of producing a fragment in particular quantum states $P(E,J \rightarrow v_1,N_1,v_2,N_2)$ using the PST are the energy E and angular momentum J of the excited molecule, the threshold energy for the decomposition E_0, and the coefficient for the attractive interaction between the fragments. (For molecules that have large attractive interactions such as hydrogen peroxide, the calculation is rather insensitive to the exact form of the attraction and the exact magnitude of the coefficient.[31,32]) The SACM contains the additional parameter α that is, in principle, adjustable. In fact, it can often be obtained from the value of the high-pressure thermal rate constant for recombination of the fragments. We use our spectroscopic model to calculate the distribution of energies and angular momenta prepared by the excitation laser $P(E_i,J_i)$ and average the probabilities given by the statistical models over this distribution for comparison with experiment:

$$n(v_1,N_1,v_2,N_2) = n_0 \sum P(E_i,J_i)P(E_i,J_i \rightarrow v_1,N_1,v_2,N_2) \qquad (4)$$

For the $6v_{OH}$ region, where the excess energy is substantial, the spectroscopic model for $P(E_i,J_i)$ makes little difference in the predicted product-state distribution, and even a thermal distribution is adequate.[31] For the $5v_{OH}$ region, however, the particular distribution of initial states has a noticeable, but not dramatic, effect.[32]

Figure 12 shows the predictions of the two statistical models along with the experimental results. Both the PST (hatched bars) and the SACM (open bars) predict qualitatively similar distributions of products following excitation in the main and combination bands of the $6v_{OH}$ region. Each shows a maximum in the region of $N = 4$—6 and predicts population up to the energetic limit of the reaction. The SACM characteristically predicts less rotational excitation than the PST. The PST seems to be in slightly better agreement with the measurements for the main band while the SACM is superior for the combination band. Both predict more rotational excitation for the higher-energy excitation of the combination band compared to the main band, as must all statistical theories, in contradiction to the measurements. This is a quantitative manifestation of the qualitatively unique behaviour of torsional excitation discussed above. Both theories reproduce the general features of the product-state distributions resulting from excitation in the $6v_{OH}$ region fairly well, but the situation is drastically different for excitation of transitions in the $5v_{OH}$ region. The SACM prediction agrees with the experimental results for both the main band and combination band quite well. The calculation finds that most of the products are formed in the lowest rotational state, in agreement with the measurements, and predicts the observed monotonic decrease of population with increasing rotational quantum number. The PST, in contrast, predicts substantially more rotational excitation than is observed experimentally and finds less population in the lowest state than in $N = 2$.

The comparison of the calculations and measurements provides a clear example of the success of the SACM in a situation where PST is inadequate. The reduction in the population of high rotational states in the SACM compared to PST is just the difference required to bring the calculation into agreement with the experiment. A crucial feature of this application of the SACM is that the parameter α is not a free parameter used in fitting the data but rather is constrained to agree with the high-pressure thermal recombination data. Because excitation of $5\nu_{OH}$ and $6\nu_{OH}$ transitions prepares the molecules with very different excess energies, α must lie in a narrow range of values to give product-state distributions that agree with the data at both energies.[32] This range of values $\alpha = 1.0 \pm 0.15$ encompasses the value from the high-pressure thermal recombination data.

The energy-selected rate measurements and product-state distribution data available for hydrogen peroxide permit a comparison with the most important predictions of statistical theories.[62] The SACM is consistent with the direct picosecond rate measurement for excitation of $5\nu_{OH}$ in a room-temperature sample,[58] the rate inferred from the linewidth for excitation of $6\nu_{OH}$ in molecules cooled in a free-jet expansion,[34] and the product-state distributions measured following excitation in both the $5\nu_{OH}$ and $6\nu_{OH}$ regions in a room-temperature sample. The distributions obtained following excitation of the combination band of the $6\nu_{OH}$ transition hint at some special behaviour of the torsional mode that is inconsistent with fully statistical behaviour. It is now possible to obtain the product-state distributions for excitation of the $6\nu_{OH}$ transition in a free-jet expansion and potentially make another detailed comparison between the experiment and theory.[63]

The two-photon excitation through the $4\nu_{OH}$ transition to a rapidly dissociating state is not only a valuable spectroscopic tool, as discussed above, but also produces unique photodissociation dynamics in hydrogen peroxide. Almost equal amounts of OH are formed in $v = 1$ and $v = 0$ in the two-photon excitation,[43] in dramatic contrast to one-photon dissociation at even higher energies,[65-68] which produces very little ($<3\%$) vibrationally excited OH product (see Chapters 1 and 4). This is an example of the product-state distribution serving as a clear indicator of rather different photodissociation dynamics. The energy added by the two photons ($\sim 27\,000\,cm^{-1}$) is substantially less than that available in the one-photon

[65] G. Ondrey, N. van Veen, and R. Bersohn, *J. Chem. Phys.*, 1983, **78**, 3732.
[66] A. Jacobs, K. Kleinermanns, H. Kuge, and J. Wolfrum, *J. Chem. Phys.*, 1983, **79**, 3162.
[67] M. P. Docker, A. Hodgson, and J. P. Simons, *Faraday Discuss. Chem. Soc.*, 1986, **82**, in press.
[68] K.-H. Gericke, F. J. Comes, S. Klee, and R. N. Dixon, *J. Chem. Phys.*, 1986, **85**, 4463.

experiments at 248 nm[65,67] and 266 nm,[68] which add about 40 000 cm^{-1} and 38 000 cm^{-1} of energy, respectively. This result points to the special role that the highly vibrationally excited intermediate state plays in the dissociation. Apparently, some of the initial vibrational excitation is preserved in the dissociation step and appears as the vibrational excitation of the products. Presently, data are only available for excitation with two photons of the same wavelength, but future two-colour experiments in both low-pressure samples and free-jet expansions should allow us to make the overall process more efficient and to unravel the details of the excitation process. This scheme is potentially applicable to a number of systems as a means of obtaining vibrational overtone spectra of cold molecules and of exploring unique regions of electronically excited potential-energy surfaces that can be accessed through the highly vibrationally excited intermediate.

5 Conclusion

Photodissociation of highly vibrationally excited molecules is an intriguing phenomenon that provides a means of testing fundamental notions about intramolecular interactions, of discovering the nature of species that are often important in photochemistry, of exploring elementary chemical processes, and of understanding molecular behaviour in an energy regime that is relevant to reactions occurring in environments as diverse as the troposphere and flames. The preparation of highly vibrationally excited molecules by laser excitation of overtone vibrations and the spectroscopic detection of the products of the resulting dissociation provide the specificity in the excitation and the resolution in the product detection that is required to unravel the behaviour of the dissociating molecule, particularly when the spectral congestion and distribution of initial energies are reduced by cooling the molecules in a supersonic expansion.

We probe the photodissociation dynamics of highly vibrationally excited molecules through vibrational overtone spectroscopy, measurements of dissociation rates, and observation of the distribution of the decomposition products among their quantum states. The spectroscopic studies reveal the importance of both homogeneous and inhomogeneous structure in vibrational overtone excitation spectra. Analysing these spectra demonstrates the role that intramolecular couplings, such as Fermi resonances, play in the spectroscopy and shows that these spectra are potentially a window on intramolecular dynamics. Vibrational overtone spectra of small molecules cooled in a supersonic expansion are particularly informative since the resolution of such transitions can greatly aid in discovering the couplings that are operative. In addition, the spectral widths of individual transitions often carry information on the dissociation lifetime of the highly vibrationally excited molecule. Direct measurements of the dissociation rate are an incisive test of theoretical descriptions, particularly when combined with the corresponding product-state distribution measurements. Product-state distributions can also signal unique electronic photodissociation dynamics, as illustrated by the extensive vibrational excitation, compared to that produced in the one-photon process, found in the products of the dissociation of hydrogen peroxide through a bound, highly vibrationally excited intermediate state. Future studies promise to provide even more detailed information on the decomposition dynamics of highly vibrationally excited molecules and on the role that such molecules play in electronic photodissociation processes.

Acknowledgements. It is a pleasure to acknowledge the contributions of many talented co-workers, B. D. Cannon, T. R. Rizzo, E. S. McGinley, C. C. Hayden, H.-R. Dübal, T. M. Ticich, L. J. Butler, and M. D. Likar, who have carried out the work on vibrational overtone induced dissociation described here. In addition, L. J. Butler and S. M. Penn critically read parts of the manuscript, and T. M. Ticich provided crucial assistance with the figures. Different aspects of this work are supported by the Office of Basic Energy Sciences of the U.S. Department of Energy and by the U.S. Army Research Office. F.F.C. is grateful to the Alexander von Humbolt Foundation for a Senior U.S. Scientist Award and to the Institut für Physikalische Chemie, Universität Göttingen, for kind hospitality during the preparation of this manuscript.

CHAPTER 7

Characterization and Uses of Ions Generated by Multi-Photon Ionization

I. POWIS

1 Introduction

Multi-photon ionization (MPI) of molecules, while still a comparatively recent addition to the chemical physicists' armoury of techniques, has nevertheless found widespread and wide-ranging uses. So much so that the generic field of applications that in some way utilize the phenomenon of MPI now far exceeds the scope of a single article such as this. Not surprisingly, however, given the great interest that MPI phenomena generate, there are various reviews currently available that treat selected aspects of the topic.[1-10]

The most notable and best established applications are undoubtedly those in which the spectroscopy of the neutral intermediate states that are encountered in resonance-enhanced multi-photon ionization (REMPI) is considered.[2,3] Here, coherent, multi-photon absorptions to a high-lying neutral state are detected by monitoring the current resulting from the subsequent ionization of the prepared state by further photon absorptions. The ionization/detection step obviously may compete with other relaxation processes, and very elegant studies of the predissociation dynamics of the

[1] P. M. Johnson, *Appl. Opt.*, 1980, **19**, 3920
[2] P. M. Johnson and C. E. Otis, *Annu. Rev. Phys. Chem.*, 1981, **32**, 139.
[3] P. M. Johnson, *Acc. Chem. Res.*, 1980, **13**, 20.
[4] M. J. van der Wiel, *J. Chim. Phys. Phys.-Chim. Biol.*, 1980, **77**, 647.
[5] V. S. Antonov and V. S. Letokhov, *Appl. Phys.*, 1981, **24**, 89.
[6] D. H. Parker, J. O. Berg, and M. A. El-Sayed in 'Advances in Laser Chemistry', ed. A. H. Zewail, Springer Ser. Chem. Phys., Vol. 3, Springer-Verlag, Berlin, 1978.
[7] E. W. Schlag and H. J. Neusser, *Acc. Chem. Res.*, 1983, **16**, 355.
[8] D. A. Gobeli, J. J. Yang, and M. A. El-Sayed in 'Advances in Multi-photon Processes and Spectroscopy', Vol. 1, ed. S. H. Lin, World Scientific Publ., Singapore, 1984.
[9] R. B. Bernstein, *J. Phys. Chem.*, 1982, **86**, 1178.
[10] M. N. R. Ashfold, *Mol. Phys.*, 1986, **58**, 1.

quantum-state-selected excited neutrals have emerged as an important application of MPI in their own right.[10]

Naturally there have also been some investigations of the generated ions and of molecular MPI *per se*.[5,7-9] These have concentrated particularly on the observed ion fragmentation (MPIF) patterns with a view to elucidation of the dynamics of photon absorption and dissociation following the initial concerted multi-photon step. Although the available data have been obtained predominantly for conditions of moderate to high laser power,* some definite ideas of how systems generally evolve during MPI have been gained. Under such conditions, however, the net photon uptake can be expected to be potentially large, with a consequent increase in the overall complexity of the ion dissociations as compared to those of the neutral intermediate. Certainly many more pathways need to be considered in attempts to understand the ion fragmentation yields.

There is perhaps a touch of irony to this hint that a so-called ionization technique has experimentally been most readily applied and refined in applications that concern the *neutral* intermediate state. Nevertheless, the knowledge gained from the aforementioned MPIF studies, when combined with insights from more recent investigations of the molecular ionization mechanism, does now lead to the view that REMPI techniques may be exploited to generate highly quantum-state-characterized ions. Essentially this can be expected, since it is an optically prepared and hence selected neutral intermediate (resonant) state that is ionized in a REMPI process, in contrast to single-photon vacuum ultraviolet ionization, where a thermal distribution of the initial (ground) state is ionized. When other features of MPI ion preparation are recognized (*e.g.* very high temporal and spatial resolution in the ion source, and high efficiency), potential applications in the fields of molecular ion spectroscopy and reaction dynamics become apparent. We therefore set out here to review our understanding of the dynamics of molecular REMPI, emphasizing the post- rather than pre-ionization processes. With this background we shall then turn to a consideration of some examples of such applications.

2   Multi-Photon Ionization

The term multi-photon ionization is quite generally applied to any system in which multiple photons are absorbed to take the system above its first ionization potential with resultant ion formation. Photoionization can hence be achieved with photon energies less than the molecular ionization potentials and consequently at more convenient wavelengths than the v.u.v. photons required for normal single-photon ionization. The mechanism of photon uptake may be a true, concerted multi-photon absorption, or it may

---

* Such terminology is of course relative. Moreover, factors such as focusing conditions, pulse duration, and coherence of the laser are of importance. By moderate to high power we simply wish to imply conditions that result in moderate to high fragmentation

be a sequential (strictly a multiple-photon) process. Often a combination of both is encountered.

The overall cross-section for ion formation can be enhanced by the existence of an excited neutral state of the molecule that is resonant with an integral number of the incident photons. Such REMPI can then be regarded as an (at least) two-step process: excitation to the intermediate state followed by further photon absorptions to raise the system above its ionization energy. A common and convenient shorthand notation for the description of REMPI processes indicates n, the number of photons required for the initial resonant step, and m, the number subsequently absorbed to ionize the molecule, as $(n + m)$. Many molecules are 'colourless' in the wavelength regions of interest in this context (visible and near u.v.) so that often $n > 1$ and the first absorption is a genuine multi-photon process. Therefore the resonant step is itself of low intrinsic probability and, it has been established, will generally become the rate-determining step for net ionization under typical experimental conditions – this is after all the basis for the accepted interpretation of MPI spectra of the neutral intermediate.[3] As a consequence the resonant step is readily characterized and n determined. However, one corollary is that the post-resonance steps may often be kinetically saturated by the high laser fluxes that drive the multi-photon transition, and so characterization of those photon absorptions that ionize and fragment the molecule is less straightforward.

It is clearly of paramount importance that, if we are to be able to use REMPI for the clean, controlled generation and photodissociation of ions, such details can in fact be unravelled in order that we can optimize and hence fully exploit useful features of the overall ionization process. In order to establish a basis upon which to build further consideration of just this point, it seems appropriate to introduce first some of the pertinent observations that have lain at the heart of much of the experimental and theoretical development of MPIF phenomena. This is the intent of the following sections.

3 Fragmentation

Laser Intensity Dependence

As well as an often pronounced dependence upon laser wavelength (arising from the resonance enhancement of MPI cross-sections) it has been widely shown that the observed ion fragmentation behaviour varies markedly with the laser photon flux in the ionization region. For example, Boesl *et al.*[11] found benzene to be ionized by irradiation at both 10^7 W cm^{-2} *and* 10^9 W cm^{-2}. Only in the latter case, however, was extensive fragmentation of the ionic products observed. Similarly, the extent of fragmentation of other

[11] U. Boesl, H. J. Neusser, and E. W. Schlag, *J. Am. Chem. Soc.*, 1981, **103**, 5058; *Chem. Phys.*, 1981, **55**, 193.

aromatic ions can be minimized with controlled laser power,[11-13] and even van der Waals species may be ionized so as to yield abundant undissociated parent ion.[14,15]

On the other hand, with high laser intensity extensive skeletal fragmentation is readily produced. In the extreme limit multiply charged ions have been detected following MPI of benzene[16] and UF_6.[17] In early work by Bernstein's group,[18] also on benzene, ion fragments as small as C^+ were produced using ~ 390 nm photons. Simple energetic considerations dictate that at least nine such photons must be absorbed for such complete fragmentation. Comparable experiments on alkyl iodides[19] similarly suggest that the absorption of at least seven photons is necessary to account for the observed fragmentation. Moreover, these estimates are only lower limits, since the partitioning of excess energy in the various possible intermediate products is not known.

Ladder Switching and Photon Absorption

The most general formulation of the problem we thus pose is as follows: there is a net uptake of several, possibly many, photons by the system. What are the elementary steps in this process? Specifically, at what stage(s) does ionization take place? When does molecular fragmentation occur? At what point(s) are the photons absorbed? Two broad categories of behaviour can be conveniently and immediately proposed: either ionization precedes any fragmentation or alternatively there is some neutral dissociation prior to photoionization.

Several examples of the latter proposition can be found. In a MPI spectroscopic investigation of methyl iodide (monitoring total ion current) Gedanken *et al.*[20] recorded the molecular bands of the $6p \leftarrow 5p\pi$ Rydberg transition of CH_3I, two-photon resonant with the incident laser light of 360—310 nm. At shorter wavelengths the broad molecular features disappear from the recorded spectrum and instead very much sharper, atomic-like features are seen. On energetic grounds these can be identified with (2 + 1) ionization of $I(^2P_{\frac{3}{2}})$ and $I^*(^2P_{\frac{1}{2}})$. This wavelength region (310—260 nm) spans the well known A-band photodissociation continuum of CH_3I. It seems plausible, therefore, that the much larger cross-section for this one-photon dissociation gives this process precedence over any multi-photon absorptions, with the resulting I atom photofragments being subsequently ionized

[12] C. T. Rettner and J. H. Brophy, *Chem. Phys.*, 1981, **56**, 53.
[13] M. Stuke, D. Sumida, and C. Wittig, *J. Phys. Chem.*, 1982, **86**, 438.
[14] H. Shinohara and N. Nishi, *Chem. Phys. Lett.*, 1982, **87**, 561.
[15] J. B. Hopkins, D. E. Powers, and R. E. Smalley, *J. Phys. Chem.*, 1981, **85**, 3739.
[16] P. Hering, A. G. M. Maaswinkel, and K. L. Kompa, *Chem. Phys. Lett.*, 1981, **83**, 222.
[17] M. Stuke and C. Wittig, *Chem. Phys. Lett.*, 1981, **81**, 168.
[18] L. Zandee and R. B. Bernstein, *J. Chem. Phys.*, 1979, **70**, 2574; *ibid.*, 1979, **71**, 1359.
[19] D. H. Parker and R. B. Bernstein, *J. Phys. Chem.*, 1982, **86**, 60.
[20] A. Gedanken, M. B. Robin, and Y. Yafet, *J. Chem. Phys.*, 1982, **76**, 4798.

by a $(2 + 1)$ MPI mechanism.[21] Closer examination of the power dependences for the mass-resolved ionization signals bears out this intuitive view.[22] At the same time some MPI signal of the fragment CH_3 can also be identified in this region. The A-band photodissociation of CH_3I is known[23] to be sufficiently rapid ($\tau < 1$ ps) for the consecutive photodissociation–photoionization to occur within the ns duration of the laser pulse.

In a related fashion it has been shown that, when recording NO_2 MPI spectra at wavelengths below 500 nm, photofragmentation of NO_2 becomes more likely than MPI and the spectrum that is actually then observed takes on the appearance of the NO spectrum.[24]

The $C_6H_6^+$ ion is found to be a prominent feature of the MPI mass spectrum of benzaldehyde.[25–29] However, this species has been thought to result likewise from the ionization of the neutral C_6H_6 photodissociation product that is formed competitively with MPI of the parent molecule. This assumption was nicely verified by analysis of the multi-photon ionization–photoelectron spectrum (MPI–PES), the $C_6H_6^+$ precursor being identified by comparison of the benzaldehyde MPI–PES with the MPI–PES of benzene.[30] A similar photoionization of neutral photodissociation products was proposed in a study of acetone MPIF.[31] Even more extensive dissociation prior to ionization has been inferred for the MPI of CCl_2F_2, again on the basis of MPI–PES and MPIF data.[32] CF^+ constitutes the main fragment ion observed, and at the same time the MPI–PES spectrum shows features reminiscent of diatomic character. A possible sequence of events is as follows: $CCl_2F_2 \rightarrow CClF_2 \rightarrow CF(X) \rightarrow CF(B) \rightarrow CF^+ + e^-$.

To the foregoing examples may be added the well established behaviour of various organometallic compounds in which the loss of ligands from the central metal atom is either completed,[33–38] or nearly so,[39,40] prior to

[21] T. Tsukiyama, B. Katz, and R. Bersohn, *Chem. Phys. Lett.*, 1986, **124**, 309.

[22] Y. Jiang, M. R. Giorgi-Arnazzi, and R. B. Bernstein, *Chem. Phys.*, 1981, **106**, 171.

[23] J. L. Knee, L. R. Khundkar, and A. H. Zewail, *J. Chem. Phys.*, 1985, **83**, 1996.

[24] R. J. S. Morrison, B. M. Rockney, and E. R. Grant, *J. Chem. Phys.*, 1981, **75**, 2643.

[25] V. S. Antonov, V. S. Letokhov, and A. N. Shibanov, *Appl. Phys.*, 1980, **22**, 293.

[26] V. S. Antonov, V. S. Letokhov, and A. N. Shibanov, *Sov. Phys.–JETP (Engl. Transl.)*, 1980, **51**, 1113.

[27] A. V. Polevoi, V. M. Katyuk, G. A. Grigor'eva, and V. K. Potapov, *Khim. Vys. Energ.*, 1984, **18**, 195.

[28] J. J. Yang, D. A. Gobeli, R. S. Pandolfi, and M. A. El-Sayed, *J. Phys. Chem.*, 1983, **87**, 2255.

[29] J. J. Yang, D. A. Gobeli, and M. A. El-Sayed, *J. Phys. Chem.*, 1985, **89**, 3426.

[30] S. R. Long, J. T. Meek, P. J. Harrington, and J. P. Reilly, *J. Chem. Phys.*, 1983, **78**, 3341.

[31] M. Baba, H. Shinohara, N. Nishi, and N. Hirota, *Chem. Phys.*, 1984, **83**, 221.

[32] J. W. Hepburn, D. J. Trevor, J. E. Pollard, D. A. Shirley, and Y. T. Lee, *J. Chem. Phys.*, 1982, **76**, 4287.

[33] S. Leutwyler, U. Even, and J. Jortner, *J. Phys. Chem.*, 1981, **85**, 3026.

[34] S. Leutwyler, U. Even, and J. Jortner, *Chem. Phys. Lett.*, 1980, **74**, 11.

[35] S. Leutwyler, U. Even, and J. Jortner, *Chem. Phys.*, 1981, **58**, 409.

[36] P. C. Engelking, *Chem. Phys. Lett.*, 1980, **74**, 207.

[37] S. Leutwyler and U. Even, *Chem. Phys. Lett.*, 1981, **84**, 188.

[38] D. P. Gerrity, L. J. Rothberg, and V. Vaida, *Chem. Phys. Lett.*, 1980, **74**, 1.

[39] G. J. Fisanick, A. Gedanken, T. S. Eichelberger, N. A. Kuebler, and M. B. Robin, *J. Chem. Phys.*, 1981, **75**, 5215.

[40] D. A. Lichtin, R. B. Bernstein, and V. Vaida, *J. Am. Chem. Soc.*, 1982, **104**, 1830.

ionization. Nevertheless, all the preceding cases seem to form in a sense exceptional, rather than normal, instances of MPI behaviour. More commonly it has been found that ionization of the parent species precedes any molecular fragmentation.

It has been the practice to discuss these mechanisms in terms of energetic 'ladders' of states up which the system is driven by photon absorption. There is thus a neutral ladder, a ladder of parent-ion states, and families of ladders for each of the potential photofragment species (ionic and neutral). Since, when mass spectrometric detection is used, it is only the charged fragments that are actually detected, one can remain quite ignorant of the neutral photofragment ladders, unless these also lead to secondary photoionization. The post-ionization mechanistic questions can be restated as whether or not ladder switching occurs (*i.e.* whether photon absorption transfers into the manifold of fragment- rather than parent-ion states) and, if so, at what point. Post-ionization steps can consequently be considered to be either multiple- or single-ion photodissociations.

Such issues can be experimentally probed by investigating the intensity dependence of the various fragments, combining the results with energetic data, or else by studying the temporal evolution of the MPI system using time-resolved two-colour experiments and studies in which the laser pulse duration is varied. The ionization step itself may be strongly characterized by photoelectron energy analysis.

Kinetic Rate Models

Many of the earlier experiments in which extensive MPI fragmentation was recorded have been analysed in terms of a rate equations model.[31,41-46] This is a conceptually appealing approach, although certain preconditions have to be met.[47] Essentially these require that the coherent steps are dominated by incoherent processes. This usually poses no practical problems, especially since the lasers employed typically have only short coherence times. Coherence effects that would invalidate this approach can, however, be found.[39]

The essence of this method consists of creating a system of linear differential equations for the time-dependent population of each plausible intermediate. To do this requires making assumptions to identify likely reaction paths and assigning an effective rate to each. This can be done in terms of an effective cross-section and laser intensity dependence for each step. The limitation is that the cross-section is generally not known.

[41] J. P. Reilly and K. L. Kompa, *J. Chem. Phys.*, 1980, **73**, 5468.
[42] G. J. Fisanick, T. S. Eichelberger, B. A. Heath, and M. B. Robin, *J. Chem. Phys.*, 1980, **72**, 5571.
[43] G. J. Fisanick and T. S. Eichelberger, *J. Chem. Phys.*, 1981, **74**, 6692.
[44] B. D. Koplitz and J. K. McVey, *J. Chem. Phys.*, 1984, **80**, 2271.
[45] B. D. Koplitz and J. K. McVey, *J. Chem. Phys.*, 1984, **81**, 4963.
[46] B. D. Koplitz and J. K. McVey, *J. Phys. Chem.*, 1985, **89**, 2761.
[47] J. R. Ackerhalt and J. H. Eberly, *Phys. Rev. A*, 1976, **14**, 1705.

Nevertheless, the system of equations can be manipulated and certain limiting conditions introduced. In such cases overall intensity dependences, I^p, may be obtained for the abundance of different fragments and can be directly compared with experiment, although again caution is required because the generally ill-defined spatial variation of light intensity at the laser focus can give rise to unrecognized saturation effects.

An investigation of acetaldehyde exemplifies this approach.[42,43] Over an extensive range of laser flux the log–log plot of product yield *versus* intensity for each fragment shows one or two linear regions of integral power dependence, separated in the second case by an intermediate region of non-integral dependence. The rate model can be refined by making allowance for possible fragment photodissociation paths, hopefully increasing thereby its physical content and reasonableness. In this fashion, fair, though not exact, agreement with experiment can be obtained.

The predictive power of the rate equations approach is necessarily limited, and so it serves best to rationalize experimental observation. This is clearly of most value in systems of limited complexity. Bromobenzene is one such relatively simple example,[44] and here a ladder-switching mechanism is strongly indicated: (a) from the neutral to the parent ion after two-photon absorption and (b) to the phenyl fragment-ion manifold after a single-photon absorption by the parent ion, this being followed by a further single-photon dissociation to the $C_4H_x^+$ ion, at which the fragmentation processes terminate.

The heuristic value of this rate equations modelling should not be minimized. It demonstrates the time dependence of fragmentation through the integrated system of equations, showing that different time evolution would be expected for a system that climbs a single ladder by successive photon absorptions as opposed to one that experiences successive photodissociations or ladder switching. Evidently time-resolved measurements should be extremely informative.

Statistical Approaches

An alternative approach to the theoretical modelling of MPI fragmentation behaviour is the application of some form of statistical treatment. The experimentally observed trends all point to an increase in the extent and number of fragmentation channels as the laser power increases and hence, one assumes, as the mean energy absorbed by a system increases. This accords with crude intuition – as we rise further above a number of energetic thresholds, dynamics will become less important and a randomness sets in, *i.e.* entropy supplants simple energetic considerations. On this premise a pure statistical theory was proposed by Silberstein and Levine.[48,49]

The essence of this approach lies in computing the final composition of a system of N parent molecules to be that of maximal entropy subject to

[48] J. Silberstein and R. D. Levine, *Chem. Phys. Lett.*, 1980, **74**, 6.
[49] N. Ohmichi, J. Silberstein, and R. D. Levine, *J. Phys. Chem.*, 1981, **85**, 3369.

imposed constraints. Initially,[48] these were considered to be minimally the conservation of charge and elemental composition and, secondly, the conservation of energy, which is taken to be characterized by a mean, $<E>$. Then it is simply the accessible product phase space that determines the relative fragment abundance. It is undoubtedly an extreme statistical proposition but one that allows qualitative explanation of observed behaviour through a computational scheme.[50] Indeed it makes some simple predictions that do not require computation for their evaluation. First, since only energy and composition enter as constraints, it should be possible to obtain similar fragmentation patterns for different isomers, presumably by finding conditions (photon wavelength and flux) under which $<E>$ is identical. This was subsequently confirmed in experiments on azulene and naphthalene[51,52] and for *cis*- and *trans*-dichloroethane.[53] Secondly, for a given molecule it should be possible to find similar fragmentation for different wavelengths, again on the assumption that a similar $<E>$ can be effected by varying laser intensity. This too can be tested.[54] Nevertheless, it is undoubtedly too simplistic to attempt to argue that a simple mean is sufficient to characterize the energy of the system. While it may be adequate for, say, electron impact ionization where excitation occurs in a single collision,[55] the sequential excitation steps encountered in MPI may prove less amenable to this approximation. The precise distribution will be dependent on the number of photons absorbed and, where relatively few are absorbed (in the u.v. and at low laser powers), will differ markedly from a uni-modal form.

A more immediate refinement is, however, the introduction of some additional, dynamical constraints; specifically one wishes to nominate the possible reaction pathways (equivalently to close some inherently unlikely ones) in a manner that will hopefully reflect the dissociation mechanisms in force. Otherwise the maximal entropy formalism can be retained.[56] To understand this approach fully it is necessary to appreciate the methodology as well as the postulates. The theory is used for simulation of the experiment, and further assumptions are systematically introduced in order to improve the agreement. In this manner conclusions as to the mechanism can be drawn. It is emphatically not a predictive method, and such dynamical constraints as are introduced come about through a deductive process rather than as initial postulates.

This can be made clear with a consideration of Silberstein and Levine's treatment of alkyl iodides.[56,57] First, all 'reasonable' parent dissociation pathways leading to stable products are assumed to be possible at their threshold, except some four-centre eliminations, which were assumed to

[50] J. Silberstein and R. D. Levine, *J. Chem. Phys.*, 1981, **75**, 5735.
[51] D. M. Lubman, R. Naaman, and R. N. Zare, *J. Chem. Phys.*, 1980, **72**, 3034.
[52] D. M. Lubman, *J. Phys. Chem.*, 1981, **85**, 3752.
[53] J. W. Hudgens, M. Seaver, and J. J. DeCorpo, *J. Phys. Chem.*, 1981, **85**, 761.
[54] D. A. Lichtin, R. B. Bernstein, and K. R. Newton, *J. Chem. Phys.*, 1981, **75**, 5728.
[55] J. Silberstein and R. D. Levine, *J. Am. Chem. Soc.*, 1985, **107**, 8283.
[56] J. Silberstein and R. D. Levine, *Chem. Phys. Lett.*, 1983, **99**, 1.
[57] J. Silberstein, N. Ohmichi, and R. D. Levine, *J. Phys. Chem.*, 1985, **89**, 5606.

have activation barriers. The relative importance of each of these channels can then be calculated from the formalism with $<E>$ serving as a parameter as before, and fragment branching ratios are then obtained. Comparison with the experimental fragmentation pattern readily established an optimum value of $<E>$. Subsequently the agreement between theory and experiment could be improved if it were assumed that stable products were able to be further photodissociated. Thus dominant secondary dissociation pathways for each stable primary ion, i, could be computed in an analogous manner and values of $<E_i>$ for each potential ion photodissociation could be deduced. In part these energies are derived from residual energy from the parent dissociation but *must* in part be acquired through further photon absorption by the otherwise stable fragment. This approach therefore allows the likelihood of ion photodissociation channels to be assessed, rather than making any firm prior assumptions as to their involvement. Clearly this stage of refinement could then be repeated for each of the secondary fragment ions, *etc.*

A principal conclusion of this simulation was that fragment-ion photodissociation is in general very likely.[57] For any given ion in the net dissociation scheme $<E_i>$ was typically not greater than 1 photon equivalent above its dissociation threshold. Consequently excited-ion states play only a small role, and the 'ladder-switching' mechanism would appear to be operative in these systems. A related observation was that, whilst the computed fragmentation patterns were very sensitive to the parameters $<E_i>$, the identity of the dominant mechanistic pathways was not. In other words, more net photon absorption drives the system further along the network of ladders rather than opening up new, highly energetic pathways from the parent. Under conditions of moderate laser flux or when the laser wavelength does not correspond to suitable transitions in the ions, little photodissociation will be possible.

An alternative, more mechanistic statistical rate equations approach has been introduced by Rebentrost *et al.*[58,59] These authors assume that all photon absorption after ionization is confined to the parent-ion species, which sequentially absorbs multiple photons, converting back to a vibrationally excited ground state prior to any dissociation taking place. Dissociations then occur via parallel and, where the fragments retain sufficient energy, consecutive pathways all originating from an energy-rich parent ion; no further photons are absorbed by fragment ions. The rates of each decomposition can then be calculated by statistical methods, specifically by a phase space formulation that also allows the energy deposition function for each fragment ion to be obtained, given an assumed energy distribution in the parent ion. This approach has been evaluated in comparisons with experimental data for benzene,[59] benzaldehyde, and phenol.[60] Quite good agreement with experimental intensity dependence data may be obtained.

[58] F. Rebentrost, K. L. Kompa, and A. Ben-Shaul, *Chem. Phys. Lett.*, 1981, **77**, 394.
[59] F. Rebentrost and A. Ben-Shaul, *J. Chem. Phys*, 1981, **74**, 3255.
[60] J. J. Yang, M. A. El-Sayed, and F. Rebentrost, *Chem. Phys.*, 1985, **96**, 1.

Nevertheless, there are grounds for questioning the physical content of this model.

Insofar as the phase space and RRKM formulations for unimolecular decay rates are essentially equivalent for most ion dissociations (due to the prevalence of loose transition states), this model is effectively the same as the established quasi-equilibrium theory (QET) for electron impact mass spectra.[61-63] Here the energy is clearly correctly considered to be deposited in the parent ion prior to any fragmentation, since there is only the single collision event to ionize/energize the system. In the MPI situation, however, the sequential absorption of energy allows, in principle, for a greater complexity of feasible reaction paths, *viz.* fragment-ion photodissociations. The assumption that all photons are absorbed by the parent ion may then be justified for computational convenience but not physical exactness. As pointed out by Rebentrost *et al.*,[59] this scheme would require approximately 60 eV of energy in the parent ion of benzene to make C^+ the most abundant fragment, as is actually observed. As such high excitation energies are approached, many unimolecular rates must become sufficiently great to permit effective competition between dissociation and further photon absorption during a typical ns duration laser pulse. Indeed, close study of the discrepancies between this theory and experiment does lead to the conclusion that multiple absorptions, *i.e.* by the fragments, must occur.[59,60] There is also convincing experimental evidence, considered below, that the fragments do absorb photons.

With these caveats in mind, Dietz *et al.*[64] were led to consider a statistical rate model that explicitly includes ladder switching. Individual rates are calculated from RRKM theory with the competition between unimolecular dissociation and photon absorption serving to limit the energy deposition in any given fragment ion. Thus reaction channels with more than one photon excess of energy above a lower threshold are excluded. In application to benzene the model works well in predicting the observed fragment abundances.

There is another aspect that can be used to evaluate the relative merits of these statistical approaches, and that is the consideration of energy distribution in the products. The above theories were all developed in the context of understanding experimental fragment abundances, particularly for the MPIF of benzene. Implicitly, however, the energy in each fragment can be computed with the same theoretical framework, and indeed has to be computed where sequential fragmentations are permitted. The methods that allow for ladder switching predict substantially less energy in the products and are therefore found to be in closer accord with experimental measurements that show relatively low translational temperatures for MPIF product ions.[65,66]

[61] C. E. Klots, *J. Phys. Chem.*, 1971, **75**, 1526.
[62] H. M. Rosenstock, M. B. Wallenstein, A. L. Wahrhaftig, and H. Eyring, *Proc. Natl. Acad. Sci. U.S.A.*, 1952, **38**, 667.
[63] W. Forst, 'Theory of Unimolecular Reactions', Academic Press, New York, 1973.
[64] W. Dietz, H. J. Neusser, U. Boesl, E. W. Schlag, and S. H. Lin, *Chem. Phys.*, 1982, **66**, 105.

Time-Resolved Studies

If the preceding discussions have suggested the likely role of secondary-ion photodissociation, it is clear that somewhat less ambiguous data than are afforded by MPIF fragment abundances alone are required for corroboration of this idea. This can be achieved by using some form of time-resolved investigation to probe the overall fragmentation mechanism.

Benzene MPIF. The benzene molecule has played a central role in the development of our current understanding of MPI, as already mentioned. From v.u.v. photoionization studies[67-69] it is known that there are four major dissociation products, each with a threshold of around 4 eV relative to the ground state $C_6H_6^+$ ion: $C_6H_5^+$, $C_6H_4^+$, $C_4H_4^+$, and $C_3H_3^+$. From REMPI experiments using visible photons in either $(2 + 1)^{18}$ or $(2 + 2)^{70}$ ionization schemes, however, far more extensive fragmentation is known to occur. Subsequently Boesl *et al.*[70] have carried out two-colour experiments for this molecule. At ~ 259 nm ionization can be achieved by a $(1 + 1)$ process resonant at the S_1 state. This leads solely to the parent $C_6H_6^+$ ion, with no smaller fragments being detected. Addition of an overlapped visible laser beam does, once again, induce extensive fragmentation, as far as $C_2H_x^+$ and $C_1H_x^+$ ions, even when this second colour beam is delayed with respect to the first. Since the visible laser alone produces no ionization, it is concluded that absorption of visible photons is necessary to produce fragmentation and that this takes place from a relatively long-lived ion state and not short-lived autoionizing Rydberg levels. The identification of an ion as the long-lived absorbing species has been confirmed by using a repelling voltage to eject charged species produced in the first laser pulse from the focal volume of the delayed visible laser.[71] As this is done, the effects of the second laser are minimized. Also pertinent are photoelectron studies of the benzene MPI at 248 nm.[72] In these experiments no evidence of energetic $(> 1$ eV) electrons was found, so it is evident that ionization occurs as soon as the ionization potential is exceeded and not from a more energetic autoionizing state.

   In a variant of the above two-colour experiment the second laser was spatially as well as temporally separated from the first and ions were selectively ejected *into* the displaced focal volume of the visible laser.[73] Photodissociation of mass-selected (by time of flight) fragment ions was thus very convincingly demonstrated.

[65] T. E. Carney and T. Baer, *J. Chem. Phys.*, 1982, **76**, 5968.
[66] J. C. Miller and R. N. Compton, *J. Chem. Phys.*, 1981, **75**, 2020.
[67] J. H. D. Eland and H. Schulte, *J. Chem. Phys.*, 1975, **62**, 3835; J. H. D. Eland, R. Frey, and B. Brehm, *Int. J. Mass Spectrom. Ion Phys.*, 1976, **21**, 209.
[68] T. Baer, G. D. Willett, D. Smith, and J. S. Phillips, *J. Chem. Phys.*, 1979, **70**, 4076.
[69] B. Andlauer and Ch. Ottinger, *J. Chem. Phys.*, 1971, **55**, 1471.
[70] U. Boesl, H. J. Neusser, and E. W. Schlag, *J. Chem. Phys.*, 1980, **72**, 4327.
[71] R. S. Pandolfi, D. A. Gobeli, and M. A. El-Sayed, *J. Phys. Chem.*, 1981, **85**, 1779.
[72] J. T. Meek, R. K. Jones, and J. P. Reilly, *J. Chem. Phys.*, 1980, **73**, 3503.
[73] U. Boesl, H. J. Neusser, and E. W. Schlag, *Chem. Phys. Lett.*, 1982, **87**, 1.

These experiments show that for benzene, at least, ionization occurs once energetically allowed and that further photon absorption takes place in the ion manifolds, including those of the fragment ions.

A different kind of time-dependent experiment involves comparing pulse duration effects. Kompa *et al.*,[16] using a ps laser pulse, again observed fragmentation of benzene, but with a single pulse this went only as far as C_4^+ and C_3^+ fragment clumps. When rapid pulse trains of the ps laser were used, much more extensive fragmentation, comparable to that previously observed with ns lasers, was achieved. The single ps pulse widths therefore establish a time-scale for absorption/fragmentation steps beyond the C_4^+/C_3^+ stage. Given that these are probably comparable to the unimolecular decay times, it suggests that the further fragmentation observed with longer duration irradiation results from secondary-ion photodissociation and that the parent does not absorb significantly beyond the energetic threshold for the C_4^+/C_3^+ fragmentations.

A clear overall picture of the MPIF of benzene emerges, in accord with the statistical calculations of Dietz.[64] These calculations suggested that, after absorption of one 259 nm photon by the benzene parent ion, dissociation to $C_6H_4^+/C_6H_5^+$ would compete with absorption of one further photon. At this level, dissociation to $C_4H_4^+$ and $C_3H_3^+$ was certain and all further photon absorption and dissociation were from these two daughter ions.

Benzaldehyde MPIF.   We have already indicated that MPI of benzaldehyde using ns pulses at ~ 260 nm proceeds through the benzene single-photon photofragment.[25-30] El-Sayed and co-workers have shown that different behaviour arises when ps duration pulses are employed.[29] Under these conditions solely parent ions having lost one H atom are detected. This would seem to indicate that with the higher peak pulse intensity of the ps duration laser the competition between further photon absorption by the resonant S_2 state of benzaldehyde and its dissociation shifts strongly in favour of the photoionization. Hence one deduces a dissociation lifetime for this S_2 neutral state between the ps and ns regimes of the two laser experiments.

When 355 nm photons are used, both ns and the lower-energy ps pulses similarly produce ions of essentially parent mass. As the relative power of both lasers is increased, more ion fragmentation sets in, but with distinct differences apparent between the ns and ps pulses. In particular the ratio of $C_4^+ : C_6^+$ fragment clumps is greater for the ps laser. A consideration[29] of the probable dissociation pathways shows that these differences can be rationalized if the more energetic routes to the $C_4H_3^+$ fragment become more likely than indirect, sequential photodissociation paths that lead to the same fragment. In other words, ladder switching into the manifold of fragment-ion states becomes less probable. It would then appear that by increasing the laser intensity with shorter duration pulses the rate of up-pumping of a state, relative to dissociation of that state, can be increased so as to allow the up-pumping to become a more significant path.

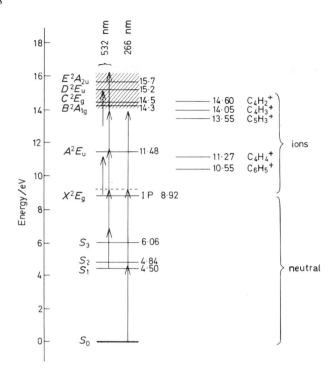

Figure 1 *Energetics of neutral and cationic 2,4-hexadiyne and appearance energies of its fragment ions*

2,4-Hexadiyne MPIF. The known energetics of the neutral, cation, and daughter ions of 2,4-hexadiyne are shown in Figure 1. It is apparent that four visible photons are required for ionization, the four-photon limit lying at 556 nm. From this threshold to ~530 nm Baer observed almost exclusively parent-ion formation with a ns laser.[74] At shorter wavelengths extensive fragmentation sets in. From the figure it is seen that at least one more photon has to be absorbed to exceed a dissociation limit of the parent ion, and the ability of the system to do this shows a marked wavelength dependence, although on either side of the apparent threshold this dependence is quite limited. Again the energy-level diagram makes it clear that it is the necessity to excite the $\tilde{A}$–$\tilde{X}$ transition that creates this marked wavelength dependence. Thus on this interpretation daughter-ion formation occurs through the first excited ion state, which can, however, be blocked by the use of an inappropriate wavelength.

El-Sayed and co-workers[75–77] have studied this same system using time-

[74] T. Carney and T. Baer, *J. Chem. Phys.*, 1981, **75**, 477.
[75] D. A. Gobeli, J. R. Morgan, R. J. St Pierre, and M. A. El-Sayed, *J. Phys. Chem.*, 1984, **88**, 178.
[76] D. A. Gobeli, J. D. Simon, and M. A. El-Sayed, *J. Phys. Chem.*, 1984, **88**, 3949.
[77] D. A. Gobeli and M. A. El-Sayed, *J. Phys. Chem.*, 1985, **89**, 1722.

resolved two-colour ps excitation sources. In their experiments ionization is achieved by a $(1 + 1)$ scheme through the S_1 manifold of the neutral using 266 nm photons. Subsequent excitation for fragmentation is assisted by absorption of delayed visible 532 nm photons (insufficiently intense to cause ionization themselves). The abundance of various fragment ions could then be investigated as a function of laser delay. Here we concentrate on the more important of these.

Since it is known from Baer's work[74] that 532 nm photons are barely adequate to excite the ground-state ion, fragmentation must depend on the absorption of a third u.v. photon taking the ion to its diffuse second excited state. At this stage $C_4H_4^+$ but not $C_4H_3^+$ fragments are energetically accessible. The effect now of a temporally coincident 532 nm laser is to deplete the former fragment in favour of the latter.[76-77] Evidently the green photon is pumping the system up out of the vibronic state that produces $C_4H_4^+$. When the delay time of the visible photon is increased, however, this effect is reduced and is no longer apparent after ~ 20 ns. It is suggested that this represents the characteristic relaxation time of the initially populated vibronic level with respect to some other state *en route* to the $C_4H_4^+$ product; possibly this would be the vibrationally excited ground state as postulated by QET theory.

The $C_6H_5^+$ fragment manifests in part the same behaviour as the $C_4H_4^+$ ion,[76-77] as might be expected given their similar threshold around the three u.v. photon level, but there is additionally some delay-time-dependent variation in this ion's intensity apparent on a 3×10^{-10} s time-scale. This is more marked at higher u.v. laser powers. It may therefore be that, as well as the competing production of $C_6H_5^+$ at the three u.v. photon level (having the 20 ns lifetime), there is also some production of this fragment via a four u.v. photon level state displaying a 350 ps lifetime.

A general observation may be made in summary of the above discussions of specific systems: although complex pathways involving sequential absorptions and dissociations need to be considered, the details of the preferred mechanisms can be deciphered by following the time evolution of the system. Moreover, by selection of laser frequency, pulse duration, and intensity, certain paths can be selectively enhanced and others can be disfavoured or closed so that control can be exerted by the experimentalist.

4 Ionization of the Resonant Intermediate

Examination of the kinetic-energy distribution of the MPI-generated photo-electrons provides the most direct means of acquiring details of the ionization step in an overall MPIF process, and it further offers the ability to characterize the ions that are formed initially. In principle, the experiment required to do this is completely analogous to conventional single v.u.v. photon photoelectron spectrometry. The pulsed nature of the MPI photon source encourages the use of time-of-flight (TOF) energy analysis as well as more conventional electrostatic deflection analysers. There are, however, disadvantages accruing from the low-duty cycle of the ionizing radiation. The total number of ions created per laser pulse must be kept low so that space charge repulsion in the highly localized ionization volume does not distort the spectrum. On the other hand, TOF analysis can offer superior resolution (~ 3 meV) under some circumstances.[78] Such technology as pulsed supersonic nozzle expansions can be conveniently incorporated in a laser MPI experiment.[79] Finally, the ability to ionize through different wavelength-selected intermediate states can dramatically enhance the versatility of a laser MPI experiment in such matters as identifying the adiabatic ionization energy and assigning vibrational progressions.[80,81] An overview of the field of MPI–PES may be found in a recent account by Kimura.[82]

We have already seen that a measurement of the mean MPI photoelectron energy can be useful to establish how high up the ladder of neutral photon absorptions the system climbs before ionization in the cases of benzene[72] and benzaldehyde.[30] Further instances include MPI of I_2[66] and H_2S.[83,84] In this latter case dissociative autoionization of a super-excited neutral, which had been postulated from MPIF studies,[85] was ruled out in favour of parent-ion formation at the lowest, four-photon, threshold.

It should be reiterated that the *n*-photon resonance excitation is assumed to be kinetically decoupled from the *m*-photon ionization step. Experimental demonstration of this decoupling can be found by two-colour experiments that are designed to excite then ionize the molecule. As an example, aniline can be excited via its S_1–S_0 0–0 transition and subsequently ionized by a 305 nm pulse.[86] As the ionizing pulse is delayed with respect to the initial excitation pulse, the net ion signal decays with a 8.9 ns time constant. As this is essentially the same as the S_1 fluorescence decay, but much less than the collisional deactivation lifetime, it suggests that ionization becomes much less probable after intersystem crossing. This can be understood and interpreted in terms of differing vertical ionization energies. That of the vibrationally excited T_1 state will likely exceed the energy available from a single 305 nm photon whereas this is sufficient for vertical ionization of the vibrationless S_1 state. Similar observations that indicate relaxation of the intermediate state have been made for benzene[87,88] and toluene.[89] A consequence of this viewpoint, which is indeed implicit in the proffered explanation of the

[78] W. G. Wilson, K. S. Viswanathan, E. Sekreta, and J. P. Reilly, *J. Phys. Chem.*, 1984, **88**, 672.

[79] J. C. Miller and R. N. Compton, *J. Chem. Phys.*, 1986, **84**, 675.

[80] S. L. Anderson, D. M. Rider, and R. N. Zare, *Chem. Phys. Lett.*, 1982, **93**, 11.

[81] J. T. Meek, E. Sekreta, W. Wilson, K. S. Viswanathan, and J. P. Reilly, *J. Chem. Phys.*, 1985, **82**, 1741.

[82] K. Kimura, *Adv. Chem. Phys.*, 1985, **60**, 161.

[83] J. C. Miller, R. N. Compton, T. E. Carney, and T. Baer, *J. Chem. Phys.*, 1982, **76**, 5648.

[84] Y. Achiba, K. Sato, K. Shobatake, and K. Kimura, *J. Chem. Phys.*, 1982, **77**, 2709.

[85] T. E. Carney and T. Baer, *J. Chem. Phys.*, 1981, **75**, 4422.

[86] D. J. Moll, G. R. Parker, jun., and A. Kuppermann, *J. Chem. Phys.*, 1984, **80**, 4808.

[87] M. A. Duncan, T. G. Dietz, M. G. Liverman, and R. E. Smalley, *J. Phys. Chem.*, 1981, **85**, 7.

[88] C. E. Otis, J. L. Knee, and P. M. Johnson, *J. Chem. Phys.*, 1983, **78**, 2091; *J. Phys. Chem.*, 1983, **87**, 2232.

[89] T. G. Dietz, M. A. Duncan, and R. E. Smalley, *J. Chem. Phys.*, 1982, **76**, 1227.

aniline experiment, is that as a first approximation multi-photon ionization should be governed by the Franck–Condon vibrational overlap between the ionic and the neutral intermediate state.

Vibrational Distribution of the Ion

Vibrational distributions of MPI-generated ions are simply inferred from the vibrational features of the corresponding MPI–PES. Amongst the larger molecules for which vibrational analyses have been undertaken are benzene,[90] chlorobenzene,[80] toluene,[91] aniline,[81,92] phenol,[93] and triethylamine.[94] For these molecules there is a definite propensity for ionization to preserve the vibrational population of the resonant intermediate in the ion, *i.e.* for $\Delta v = 0$ ionizing transitions. In general terms this may be simply rationalized in terms of favourable Franck–Condon factors between the ion and the resonant intermediate. Since the vibrational state of the intermediate is optically selected by tuning the multi-photon excitation from the ground state, there is an intriguing prospect of one being able to transfer a predetermined vibrational quantum number from the prepared intermediate through to the ion, thereby achieving vibrationally state-selective ionization.

This has in fact been well demonstrated for a number of small molecules including the preparation of $N_2^+(A^2\Pi_u)(v = 1$ or $2)$ through the $^1\Pi_u(v = 1$ or $2)$ intermediate,[95] ionization of H_2 through the $C^1\Pi_u$ $(v = 0\text{—}4)$,[96] and $B^1\Sigma_u^+(v = 0)$[97] intermediates as well as some $(3 + 1)$ ionizations of NO and NH_3.[98]

Undoubtedly the optimum circumstances for the observation of Franck–Condon-governed $\Delta v = 0$ state-selective MPI are that the intermediate should be a Rydberg state sharing the same core configuration as the ion to be formed and that it should be an $(m + 1)$ process. The Rydberg state is thus expected to have essentially the same geometry as the ion, favouring vertical ionization, and the actual ionization step is a 'clean' one-electron excitation from the outer Rydberg orbital. Viewing such MPI processes as one-photon ionizations of an optically prepared, aligned state provides a suitable theoretical basis for developing an understanding of multi-photon ionization dynamics.[99]

[90] Y. Achiba, K. Sato, K. Shobatake, and K. Kimura, *J. Chem. Phys.*, 1983, **79**, 5213.
[91] J. T. Meek, S. R. Long, and J. P. Reilly, *J. Phys. Chem.*, 1982, **86**, 2809.
[92] M. A. Smith, J. W. Hager, and S. C. Wallace, *J. Chem. Phys.*, 1984, **80**, 3097.
[93] S. Anderson, L. Goodman, K. Krogh-Jespersen, A. G. Ozkabak, R. N. Zare, and C. Zheng, *J. Chem. Phys.*, 1985, **82**, 5329.
[94] M. Kawasaki, K. Kasatani, H. Sato, Y. Achiba, K. Sato, and K. Kimura, *Chem. Phys. Lett.*, 1985, **114**, 473.
[95] S. T. Pratt, P. M. Dehmer, and J. L. Dehmer, *J. Chem. Phys.*, 1984, **80**, 1706.
[96] S. T. Pratt, P. M. Dehmer, and J. L. Dehmer, *Chem. Phys. Lett.*, 1984, **105**, 28.
[97] H. Rottke and K. H. Welge, *Chem. Phys. Lett.*, 1983, **99**, 456.
[98] Y. Achiba, K. Sato, K. Shobatake, and K. Kimura, *J. Chem. Phys.*, 1983, **78**, 5474.
[99] S. N. Dixit and V. McKoy, *J. Chem. Phys.*, 1985, **82**, 3546.

One-Photon Ionization

Ammonia. Zare and co-workers[100,101] have examined the MPI–PES of jet-cooled NH_3 by $(2 + 1)$ REMPI through various vibrational levels of the $\tilde{B}$ and $\tilde{C}'$ states. These intermediates are, respectively, $3p$ Rydbergs of E'' and A_1' symmetry converging to the first ionization limit. Figure 2 shows the pronounced state-selective $\Delta v = 0$ nature of the ionization through the vibrationless level of the $\tilde{C}'$ state. Results for the other intermediates are summarized in Table 1, from which it is seen that the 'contamination' by $\Delta v \neq 0$ transitions is never worse than 20% for the $\tilde{C}'$ state intermediate and never worse than 30% for the $\tilde{B}$ intermediate.

Table 1 $NH_3^+(\tilde{X})$ *vibrational distributions following ionization from different vibrational levels of the intermediate state,* $NH_3(v_2)^a$

Intermediate state	$\Delta v = 0$	Vibrational change, Δv_2 $\Delta v = -1$	Other
$\tilde{B}(3)$	0.83	0.06	0.11
$\tilde{B}(4)$	0.82	0.04	0.14
$\tilde{B}(5)$	0.78	0.04	0.18
$\tilde{B}(7)$	0.76	0.11	0.13
$\tilde{B}(8)$	0.74	0.10	0.16
$\tilde{B}(9)$	0.74	0.11	0.15
$\tilde{B}(10)$	0.74	0.10	0.16
$\tilde{C}'(0)$	1.0	—	—
$\tilde{C}'(2)$	0.94	0.04	0.02
$\tilde{C}'(3)$	0.83	0.05	0.12
$\tilde{C}'(4)$	0.80	0.05	0.14
$\tilde{C}'(5)$	0.80	0.05	0.14
$\tilde{C}'(6)$	0.88	0.06	0.06

a W. E. Conaway, R. J. S. Morrison, and R. N. Zare, *Chem. Phys. Lett.*, 1985, **113**, 429

Pallix and Colson[102] have closely examined $(2 + 1)$ MPI through selected near-degenerate levels of the $\tilde{B}$, $\tilde{C}'$, and $\tilde{D}$ states. They assign a small peak in the MPI–PES via the $\tilde{C}'(v_2 = 6)$ level, which is additional to the principal $\Delta v = 0$ peak as due to ion formation with $v_2 = 10$. Likewise, a small additional peak in the $B(v_2 = 10)$ MPI–PES is attributed to production of $v_2 = 5$ ion. It is inferred that there is a small coupling between the rovibronic manifolds of the $\tilde{B}$ and $\tilde{C}'$ states. Ionization from the $\tilde{D}(v_2 = 0)$ state shows $\ll 1\%$ $\Delta v = 0$ transition; rather the ion is formed with mainly $v_2 = 6$ and, to a lesser extent, $v_2 = 10$ vibrational excitations. After finding the same behaviour with a ps laser, it was concluded that there is rapid (< 5 ps) internal conversion of the $\tilde{D}$ to the $\tilde{C}'(\tilde{B})$ states.[102]

[100] W. E. Conaway, R. J. S. Morrison, and R. N. Zare, *Chem. Phys. Lett.*, 1985, **113**, 429.
[101] R. J. S. Morrison, W. E. Conaway, T. Ebata, and R. N. Zare, *J. Chem. Phys.*, 1986, **84**, 5527.
[102] J. B. Pallix and S. D. Colson, *J. Phys. Chem.*, 1986, **90**, 1499.

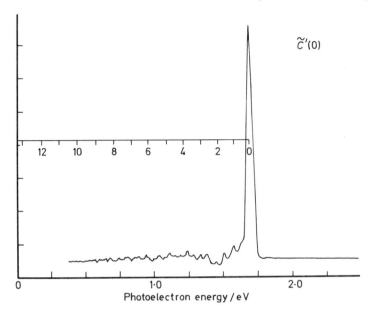

Figure 2 (2 + 1) *MPI–PES of ammonia produced via the vibrationless level of the $\tilde{C}'$ resonant intermediate state. The corresponding v_2 vibrational levels of the ground-state ion are indicated*
(Reproduced with permission from *J. Chem. Phys.*, 1986, **84**, 5527)

Earlier work had considered (3 + 1) MPI–PES of ammonia. Achiba *et al.*[98] likewise observed strong $\Delta v = 0$ ionization of the $\tilde{C}'$ state; so, too, did Glownia *et al.*,[103] but additionally these authors found a strong peak for electrons of near-zero kinetic energy, signifying NH_3^+ ions formed with maximum vibrational excitation. This particular manifestation of deviation from the simple direct Franck–Condon ionization model is not unusual, but, although it is usually taken to be indicative of some alternative ionization mechanism, it is not always easy to interpret, as indeed in the present case.

Methyl Iodide.   Woodward *et al.*[104] have reported on the (2 + 1) REMPI of methyl iodide via two $6s$ Rydberg states having A and E symmetries. Both converge to the excited $^2E_{\frac{1}{2}}$ state of the CH_3I^+ molecular ion and so deviate from our postulated optimum case. However, it seems likely that a single configuration does not adequately describe these ion core states, as attested by the extensive (single-photon) autoionization seen between the $^2E_{\frac{3}{2}}$ and $^2E_{\frac{1}{2}}$ ionization limits.[105] It is also noted that these states may have some valence character.[104]

103 J. H. Glownia, S. J. Riley, S. D. Colson, J. C. Miller, and R. N. Compton, *J. Chem. Phys.*, 1982, **77**, 68.
104 A. W. Woodward, S. D. Colson, W. A. Chupka, and M. G. White, *J. Phys. Chem.*, 1986, **90**, 274.
105 D. M. Mintz and T. Baer, *J. Chem. Phys.*, 1976, **65**, 2407.

At the wavelengths used for two-photon access to the vibrationless levels of these states the third photon is barely able on energetic grounds to produce the excited spin–orbit state of CH_3I^+, and the MPI–PES shows vibrationless $CH_3I^+(^2E_{\frac{3}{2}})$ as the predominant ionization product with a lesser population of $v_2 = 1$ (and for the E-state intermediate also some $v_2 = 2$); in other words, the $\Delta v = 0$ propensity is upheld.

A $(2 + 1)$ ionization of these Rydbergs when they have one quantum in their v_2 vibration *is* energetically capable of producing the excited $CH_3I^+(^2E_{\frac{1}{2}})$, although only for the Rydberg intermediate of E symmetry could the vibrationally excited $v_2 = 1$ level of the $^2E_{\frac{1}{2}}$ ion be reached. Nevertheless the MPI–PES apparently show that both vibrationally excited Rydberg states ionize to the lower-energy $^2E_{\frac{3}{2}}$ state with, again, a narrow distribution of ion vibrational states in which the $\Delta v = 0$ transition is the most prominent. Hence, despite differences between the Rydberg core and the ground-state ion configurations, the simple Franck–Condon overlap argument appears to retain some validity. There is other evidence to suggest that some $CH_3I^+(^2E_{\frac{1}{2}})$ formation also takes place, even though the corresponding electrons are not evident in the MPI–PES.[104,106]

Molecular Hydrogen. The attraction of studying MPI of the H_2 molecule is of course the ability to undertake more quantitative comparison of experiment with soundly based theory. Hence, while the one-photon ionizations of the resonant C,[96] B,[107] and the interesting (E,F) double-minimum states[108,109] all qualitatively concur with one's expectations of a narrow vibrational distribution in the ion centred around the $\Delta v = 0$ transition, exact agreement with computed Franck–Condon vibrational overlap factors is not found. In part these discrepancies may result from instrumental uncertainties (transmission functions, angular acceptance, *etc.*), but it seems probable that they also convey some information regarding the photoionization dynamics.

Just what this information may be has been considered at length for the ionization via the $C^1\Pi_u(v = 0$—$4)$ levels.[96] It seems unlikely that there are any accidental resonances or perturbations to account for the discrepancies that are observed. Rather, these may reflect some marked variations of the electronic matrix element in the transition moment. Normally of course these terms are assumed constant in the simple Franck–Condon model, with the relative intensities of vibrational transitions being ascribed solely to the vibrational overlap integrals.

H_2 inevitably will figure prominently in our developing understanding of the dynamics of multi-photon ionization. Although autoionization has not been invoked in the preceding, it is nevertheless known to be a significant process in single-photon ionization, and in general one expects to encounter

[106] W. A. Chupka, A. M. Woodward, S. D. Colson, and M. G. White, *J. Chem. Phys.*, 1985, **82**, 4880.
[107] S. T. Pratt, P. M. Dehmer, and J. L. Dehmer, *J. Chem. Phys.*, 1983, **78**, 4315.
[108] E. E. Marinero, R. Vasudev, and R. N. Zare, *J. Chem. Phys.*, 1983, **78**, 692.
[109] S. L. Anderson, G. D. Kubiak, and R. N. Zare, *Chem. Phys. Lett.*, 1984, **105**, 22.

it in the circumstances of MPI also. Already theoretical treatments of autoionization in H_2 MPI have been prepared.[110-112]

Nitric Oxide. The MPI–PES of NO has been extensively studied over a number of years.[78,79,98,113-119] A variety of resonant ionization schemes have been considered. One-photon ionization of resonantly excited vibrational levels of the E,[79] F, $H(H')$,[79,98] and N[79] Rydberg states leads either predominantly or exclusively to a single $\Delta v = 0$ vibrational level of the ion as simply anticipated. Most recently Reilly and co-workers[118] have examined $(1 + 1)$ ionization through the $A^2\Sigma^+(v = 0)$ Rydberg state converging to the first ionization limit. In addition to fast electrons from the expected $\Delta v = 0$ ionization there can appear in the MPI–PES a peak of slow (near-zero kinetic energy) electrons that would correspond to the $v = 6$ state of NO^+.[79,118] Two further observations help suggest the possible origin of the slow electron peak: (i) the laser excitation spectra for the $v = 6$ and $v = 0$ peaks are different, the latter displaying more extensive rotational features of the $A-X$ transition, and (ii) the angular distribution of the fast electrons is anisotropic whereas that of the slow electrons is apparently isotropic.[118] Different mechanisms thus seem to be implicated in the production of the fast and slow electrons. The former is certainly direct ionization of the intermediate: by tuning the laser to a rotational bandhead below the NO^+ $v = 6$ threshold pure $\Delta v = 0$ ionization is observed. The slow electrons on the other hand most likely result from an accidentally resonant autoionizing state in the ionization continuum. It may be remarked that only a single ($v = 6$) vibrational peak results, and this could be indicative of the $\Delta v = -1$ propensity rule associated with a vibrational autoionization.

Electronic autoionization is necessarily involved when ground-state ions are generated from the two-photon resonant B valence state of NO because the electronic configurations differ by more than one electron.[117] Here, too, analysis of the data suggests that accidental resonances can affect the basic ionization mechanism so modifying the ion vibrational distribution.

Multiple-Photon Ionization

In comparison with $(n + 1)$-type schemes MPI with more than a single-photon ionization step shows, perhaps not surprisingly, greater deviations from a simple Franck–Condon model.

110 K. R. Dastidar and P. Lambropoulos, *Chem. Phys. Lett.*, 1982, **93**, 273.
111 K. R. Dastidar and P. Lambropoulos, *Phys. Rev. A*, 1984, **29**, 183.
112 K. R. Dastidar, *Chem. Phys. Lett.*, 1983, **101**, 255.
113 J. C. Miller and R. N. Compton, *J. Chem. Phys.*, 1981, **75**, 22.
114 J. C. Miller and R. N. Compton, *Chem. Phys. Lett.*, 1982, **93**, 453.
115 J. Kimman, P. Kruit, and M. J. van der Wiel, *Chem. Phys. Lett.*, 1982, **88**, 576.
116 M. G. White, W. A. Chupka, M. Seaver, A. Woodward, and S. D. Colson, *J. Chem. Phys.*, 1984, **80**, 678.
117 Y. Achiba, K. Sato, and K. Kimura, *J. Chem. Phys.*, 1985, **82**, 3959.
118 K. S. Viswanathan, E. Sekreta, and J. P. Reilly, *J. Phys. Chem.*, 1986, **90**, 5658.
119 K. S. Viswanathan, E. Sekreta, E. R. Davidson, and J. P. Reilly, *J. Phys. Chem.*, 1986, **90**, 5078.

Nitric Oxide. Great attention has been paid to the $(2 + 2)$ MPI of the A state:

$$NO(X^2\Pi) \xrightarrow{2h\nu} NO(A^2\Sigma^+),\nu' \xrightarrow{2h\nu} NO^+(X^1\Sigma^+),\nu$$

The observations can be summarized as follows: in all cases a strong $\Delta\nu = 0$ peak of fast electrons was seen as a result of direct, Franck–Condon ionization. Additionally there arises a peak of slow electrons ($\nu = 6$ for the $\nu' = 0$ intermediate or $\nu = 8$ for the $\nu' = 1$ intermediate).[79,113–116] Finally, a number of lesser peaks are observed for the intervening vibrational levels of the ion, which, however, show a marked sensitivity to where within the band contour associated with the resonant intermediate state the excitation takes place.[79,114,115] An example of the MPI–PES spectra is shown in Figure 3.

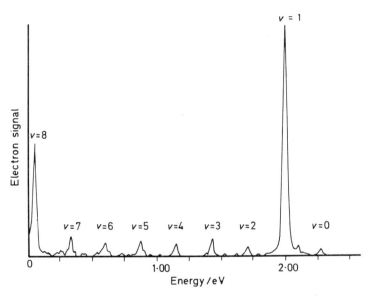

Figure 3 *(2 + 2) MPI–PES of NO produced via the $A^2\Sigma^+(\nu = 1)$ resonant intermediate state. The corresponding ion vibrational levels are indicated*
(Reproduced with permission from *Chem. Phys. Lett.*, 1982, **88**, 576)

These 'extra' photoelectron peaks have been discussed in terms of a variety of mechanistic explanations, but a consensus has emerged. Consider first the slow-electron peak. This is found to be suppressed if the resonant state is ionized by a single photon of shorter wavelength.[114] It also displays similar excitation spectra to the fast, directly ionized electrons.[115] These facts are consistent with a further, accidental resonance at the three-photon level, opening up an alternative ionization channel. A necessary feature of the new channel would be the involvement of states having some valence character so that there are alternative Franck–Condon factors brought into play that account for the apparent $\Delta v \neq 0$ behaviour between the two-photon excited Rydberg A state and the ion. Thus it may be that mixed Rydberg–valence states are populated and directly ionized; this has been seen in MPI of the B-state levels.[116,117] Alternatively, if pure valence states are accessed, they must then autoionize, since direct ionization to the ground-state ion cannot be achieved by a one-electron transition.

Identification of states at the three-photon energy (and hence identification of the ionization mechanism) is not easy because of their apparently sequential, non-coherent excitation and spectral congestion in the dense manifold of states around the A state. Using supersonic cooling, Miller and Compton[79] were able to isolate a three-photon excitation to the $B^2\Pi_{\frac{3}{2}}$ ($v = 31$) level at a wavelength immediately adjacent to the two-photon excitation from the lowest rovibronic level of NO to the $A^2\Sigma^+$ ($v = 1$) state. Excitation through this extra B-state resonance produces a MPI–PES spectrum in which *only* the slow electrons are observed. Although this is a valence state, it could presumably acquire some Rydberg character through interaction with the nearby $L(v = 9)$ and/or $C(v = 8)$ states: either would enable direct ionization to the observed $v = 8$ level of the ion. By implication, related 'accidental' resonance mechanisms may account for the electrons of near-zero energy apparent in the A-state MPI–PES.

Somewhat analogously White *et al.*[116] have proposed that at a three-photon energy close to 66 566 cm^{-1} [equivalent to the two-photon energy of the $A(v = 0)$ level] the $K(v = 1)$ state could be populated. Direct ionization of this Rydberg state would then account for a prominent $v = 1$ peak observed in the MPI–PES at this point. Moreover, this K level has mixed in some valence B ($v = 25$) character that could be expected to lead to Franck–Condon factors favouring the broader distribution of vibrational levels that is also seen. The authors note, however, that equally a four-photon resonant electronic autoionizing state could be the explanation for such a distribution.[79]

Previously it had been considered that the occurrence of pronounced single vibrational peaks other than the $\Delta v = 0$ one could be indicative of a vibrational autoionization mechanism.[113,114] This could favour a high degree of vibrational specificity by virtue of the $\Delta v = -1$ propensity rule. At the present time it remains unclear how generally important an alternative to direct ionization this might be compared to the occurrence of accidental resonances such as detailed above.

Carbon Monoxide. Despite complications, the Franck–Condon principle provides a suitable framework within which to consider MPI–PES in the preceding molecules. It is, however, less obviously useful in the case of (3 + 3) MPI via the $A^1\Pi$ state of CO.[120] The MPI–PES obtained at the $A(v = 1)$ bandhead shows a strong $v = 0$ peak for the ground-state ion, with a progression of modulated intensity stretching up to the highest energetically

[120] S. T. Pratt, E. D. Poliakoff, P. M. Dehmer, and J. L. Dehmer, *J. Chem. Phys.*, 1983, **78**, 65.

available vibrational level, $v = 9$. The spectrum at the intermediate $v = 2$ bandhead is similar, with the addition of an even more pronounced peak corresponding to formation of an excited $A^2\Pi(v = 0)$ ion. This formally requires a two-electron transition. At the $v = 3$ intermediate level a dramatic change takes place as only $(3 + 2)$ MPI to the vibrationless ground-state ion has any significant probability. This behaviour cannot be rationalized in terms of calculated Franck–Condon factors. Similar observations hold for $(2 + 2)$ MPI–PES of the same levels.[121]

No clear understanding of this system has yet emerged. It seems inevitable that all of the phenomena so far considered – accidental resonances, perturbing levels, and autoionization processes – will need to be investigated.

Rotational Distributions

As in any phototransition the change in rotational state of the ion as compared to the neutral will be subject to constraints imposed by conservation of angular momentum, parity, *etc.* Given our assumption of MPI as a decoupled excitation and subsequent ionization, there is, at least in principle, no difference in this respect between a single-photon ionization from the ground state and a one-photon ionization of the excited intermediate, a view sustained by theoretical analysis.[122] However, there may be a very significant difference in practice because of the attributes of the states being ionized: in the single-photon case these will be a thermal rotational distribution of the ground electronic state whereas in REMPI the intermediate being ionized will often be a precisely defined rovibronic level. Consequently, the rotational distribution of the ion can be expected to be constrained to be narrow in accordance with the appropriate selection rules.

Although rotational distributions of the ion might be inferred from photoelectron spectra, much as vibrational distributions are, this has only been achieved in single-photon ionization for H_2, because of limited resolution.[123–126] MPI–PES can achieve better resolution by using TOF electron energy analysis, and this has permitted both NO[18,119] as well as H_2[107] to be studied. Clearly this is a far from extensive list, yet it does allow rotational propensity rules to be evaluated. Extra information can be obtained relating to ionization of states of various symmetries by the simple expedient of tuning the laser to access different intermediate states.

Homonuclear Diatomics: H_2. The ionization of ground-state H_2 can be represented as:

$$H_2(X^1\Sigma_g^+), v'', J'' \xrightarrow{h\nu} H_2^+(^2\Sigma_g^+), v', N' + e^-(l) \qquad (1)$$

[121] S. T. Pratt, P. M. Dehmer, and J. L. Dehmer, *J. Chem. Phys.*, 1983, **79**, 3234.
[122] S. N. Dixit, D. L. Lynch, V. McKoy, and W. M. Huo, *Phys. Rev. A*, 1985, **29**, 1267.
[123] L. Åsbrink, *Chem. Phys. Lett.*, 1970, **7**, 549.
[124] Y. Morioka, S. Hara, and M. Nakamura, *Phys. Rev. A*, 1980, **22**, 177.
[125] J. E. Pollard, D. J. Trevor, J. E. Reutt, Y. T. Lee, and D. A. Shirley, *Chem. Phys. Lett.*, 1982, **88**, 434; *J. Chem. Phys.*, 1982, **77**, 34.
[126] M. W. Ruf, T. Bregel, and H. Hotop, *J. Phys. B*, 1983, **16**, 1549.

where J is the total angular momentum quantum number, N is that excluding spin, and l is the angular momentum of the continuum electron. Likewise, ionization of the B state populated in $(3 + 1)$ REMPI is:

$$H_2(B^1\Sigma_u^+), v'', J'' \xrightarrow{h\nu} H_2^+(^2\Sigma_g^+), v', N' + e^-(l) \qquad (2)$$

Pratt *et al.*[107] have shown how effective neutral-ion rotational selection rules may be simply deduced, recognizing that the final state in the ionization consists of ion plus continuum electron[127] loosely coupled in a Hund's case (d) approximation. Most notably the fundamental parity selection rule, u ⟷ g, restricts the partial wave expansion of the continuum function to odd l values in case (1) and even l values in case (2). Further, such expansions can usually be truncated after the first few terms: effectively this is to say that $\Delta l = \pm 1$ for the ejected electron in this case. Consideration of other remaining selection rules (*viz.* $a \not\longleftrightarrow s$, $+ \longleftrightarrow -$, $\Delta J = 0$ or ± 1) leads to the rotational constraints $\Delta N = \pm 1$ for the $l = 0$ s-wave electron and $\Delta N = \pm 1$ or ± 3 for the $l = 2$ d-wave electron in the MPI example, (2). Summarizing, we have a $\Delta N = \pm 1$ or ± 3 propensity rule in this case. For ground-state ionization the analogous $\Delta N = 0$ or ± 2 propensity rules may be anticipated.[128] More generally Dixit *et al.*,[122] by considering the matrix elements appearing in the theoretical cross-section expression, have demonstrated that for a Σ(neutral) to Σ(ion) one-photon ionization a '$\Delta N + l =$ odd' selection rule can be applied.

Experimental data for the ground-state ionization strongly support the $\Delta N = 0$ or ± 2 propensity rule,[123-126] and the photoelectron asymmetry in the $\Delta N = 2$ transitions has been investigated in detail.[125,129,130] Ionization of the $B(v = 6$ or $7)$ levels has also been studied.[107] In this case the MPI–PES shows clear evidence for the occurrence of $\Delta N = \pm 1$ transitions, as expected. However, the $\Delta N = \pm 3$ transitions, if present at all, are extremely weak. In fact detailed cross-section calculations indicate that these transitions are attenuated by several orders of magnitude because of interference between the d-wave components of the σ and π ionization channels.[131] Thus this REMPI scheme produces ions in one of two rotational states that are determined by the rovibronic level of the three-photon resonant intermediate state of the neutral molecule.

Heteronuclear Diatomics: NO.   An important difference may be expected for a heteronuclear, non-centrosymmetric molecule inasmuch as parity is not defined. Consequently, the partial wave expansion for the continuum electron will be of mixed parity and the constraints on the ion rotational states arising from ionization of a given rovibronic intermediate can be expected to

[127] R. A. Bonham and M. L. Lively, *Phys. Rev. A*, 1984, **29**, 1224.
[128] J. M. Sichel, *Mol. Phys.*, 1970, **18**, 95.
[129] Y. Itikawa, *Chem. Phys.*, 1979, **37**, 401.
[130] M. Raoult, Ch. Jungen, and D. Dill, *J. Chim. Phys. Phys.-Chim. Biol.*, 1980, **77**, 599.
[131] D. L. Lynch, S. N. Dixit, and V. McKoy, *Chem. Phys. Lett.*, 1986, **123**, 315.

be somewhat relaxed. However, experimental $(1 + 1)$ MPI–PES of NO via the $A^2\Sigma^+(v = 0)$ state show a clear $\Delta N = 0$ or ± 2 propensity:[78] the $\Delta N = \pm 1$ transitions were not expected to be fully resolved, but if present were much less intense.

Recalling the $\Delta N + l =$ odd selection rule enunciated by Dixit and co-workers,[122] it can be seen that this observation suggests that, notwithstanding the above comments regarding mixed parity, only odd l terms appear in the partial wave expansion. When these authors carried out a full calculation of the rotational branching for this ionization, they in fact predicted a result with a small $\Delta N = \pm 1$ contribution, which would not be inconsistent with the experimental data, given its resolution.[122] The explanation for the reduced probability of the $\Delta N = \pm 1$ ionization is found by recognizing the Rydberg nature of the orbital being ionized. This is little influenced by the anisotropy of the ionic core and in a single-centre expansion displays mainly s character, rather less d character, and very much less p character, *i.e.* it is predominantly *gerade*.[119,122] It should thus be possible to adopt usefully a $\Delta l = \pm 1$ selection rule for the ionized electron, so restricting the terms in the partial wave expansion of the continuum function.

By assuming such atomic-like character for the ionizing Rydberg orbitals, Reilly *et al.*[119] have adapted the scheme mentioned above for the derivation of H_2 photoionization propensity rules[107] to treat the MPI of NO through its A, C, and D Rydberg states. In this manner the $\Delta N = 0$ or ± 2 propensity rule observed for the A-state ionization is confirmed.[78,119] For the $C^2\Pi$ state either $\Delta N = 0$ or ± 2 or $\Delta N = \pm 1$ or ± 3 for the different Λ-doublet components is predicted. Again this was experimentally verified.[119] Finally the D state, a $3p$ $^2\Sigma^+$ state, was found experimentally to undergo $\Delta N = \pm 1$ or ± 3 transitions as predicted. Additionally, however, a large $\Delta N = 0$ contribution was observed, possibly indicating the presence of some p character in the continuum. Evidently in this case at least the mixed nature of the partial wave expansion needs to be acknowledged.[119]

5 Application Areas

From the preceding discussion of the photon absorption, ionization, and fragmentation processes encountered in a range of REMPI phenomena it is clear that no simple, universal prescriptions describing the resulting ion yield can sensibly be formulated. Nevertheless, there is ample scope for exploiting features that can be identified in a given system in order to obtain a well characterized ion sample. It is perhaps inevitable that such applications of REMPI have been slower to mature than some others, but it seems likely that much will be made of the potential of state-selective ion preparation now that the foundations of understanding and theory are in place. In order to underline this promise for the future, three current examples that elegantly utilize REMPI techniques are considered below.

Unimolecular Dissociation Rates

A traditionally profitable area for the investigation of unimolecular kinetics has been the study of metastable ions, *viz.* those ions generated in a mass spectrometer that have a lifetime comparable to the instrument residence time. Under these circumstances absolute decay rates of the order of $10^5 \, s^{-1}$ can be determined. One uncertainty inherent in the conventional experiment has been lack of a precise knowledge of the internal-energy content of the metastable-ion population; this is of enormous interest for allowing comparison with RRKM and QET theories, *etc.* Some improvement on this situation can be afforded by utilizing photoelectron–photoion coincidence (PEPICO) techniques for ion detection,[132] but there are then associated problems of instrumental sensitivity and mass resolution.

The first observation of metastable ions generated by REMPI ionization was reported by Proch *et al.*[133] These authors examined aniline ions generated by a $(1 + 1)$ REMPI process around 280 nm. They observed an asymmetric broadening of the $m/e = 66$ ion TOF peak that can be attributed to metastable dissociation:

$$C_6H_7N^+ \rightarrow C_5H_6^+ + HNC \; (k \simeq 2 \times 10^6 \, s^{-1})$$

On energetic grounds it appears that the dissociation follows absorption of a third photon after ionization. This is consistent with PEPICO results for this dissociation.[134] Eight further metastable decay channels have since been identified.[135] However, in these preliminary REMPI experiments the aniline-ion energy prior to its metastable photodissociation was not well defined. Separate MPI–PES studies[81] show that the aniline ion is formed in its ground electronic state, but, at least near the 285 nm onset for metastable dissociation,[134] there is a broad distribution of ion vibrational energy some several hundred meV wide.

Rather more promising from the point of view of state selection is the metastable reaction:

$$C_6H_5Cl^+ \rightarrow C_6H_5^+ + Cl$$

MPI–PES investigation[80,136] of the chlorobenzene molecule reveals that the ground-state ion is formed after absorption of just two photons in the 265—270 nm wavelength range, via a one-photon resonance with the $\tilde{A}^1B_2$ state of the neutral. The two-photon energy exceeds the molecular ionization potential by only $\sim 250 \, meV$, and indeed the PES show an internal-energy distribution for the ion of less than 150 meV width.[136] The ground-state ion

[132] T. Baer in 'Gas Phase Ion Chemistry', Vol. 1, ed. M. T. Bowers, Academic Press, New York, 1979.
[133] D. Proch, D. M. Rider, and R. N. Zare, *Chem. Phys. Lett.*, 1981, **81**, 430.
[134] T. Baer and T. E. Carney, *J. Chem. Phys.*, 1982, **76**, 1304.
[135] K. Kühlewind, H. J. Neusser, and E. W. Schlag, *J. Chem. Phys.*, 1985, **82**, 5452.
[136] J. L. Durant, D. M. Rider, S. L. Anderson, F. D. Proch, and R. N. Zare, *J. Chem. Phys.*, 1984, **80**, 1817.

must then absorb a further photon to be able to dissociate, although having done so it seems likely to convert back internally to the vibrationally excited electronic ground state prior to its actual fragmentation. Durant *et al.*[136] were able to measure the metastable TOF peaks over a range of wavelengths/ion energies and to infer the unimolecular rates, $k(E)$. Once again these were found to be in agreement with single-photon PEPICO data.[137,138] Thus the agreement with PEPICO data for these two molecules lends support to the general approach that has been adopted for the interpretation of REMPI ion dissociations.

The most extensively investigated system of metastable-ion dissociations observed using REMPI techniques is undoubtedly that of the benzene molecular ion.[46,139–143] The energetics for this system are indicated in Figure 4. It is seen that ground-state $C_6H_6^+(\tilde{X}^2E_{1g})$ is produced by a $(1 + 1)$ process at a wavelength of 260 nm, utilizing the S_1 intermediate state. Neusser, Schlag, and co-workers[139] have developed a novel variant of the ion TOF technique, using the so-called Reflectron instrument, which permits sensitive and diagnostic metastable-ion measurements to be made under conditions of exceptionally high mass resolution. Exciting the S_1-S_0 6_0^1 transition at 259 nm, these authors made the first REMPI observation of the following metastable reactions:[139]

$$C_6H_6^+ \rightarrow C_6H_5^+ + H \tag{3a}$$

$$C_6H_6^+ \rightarrow C_6H_4^+ + H_2 \tag{3b}$$

At the two-photon energy the parent ions can have less than 0.3 eV excess of energy; dissociation must therefore follow the absorption of further photons, although any more than one would be expected to result in a fast, not metastable, fragmentation. It is evident from the experimentally identical intensity dependence of (3a) and (3b) that they proceed from a common precursor state.[46] Subsequently, the metastable fragmentations:

$$C_6H_6^+ \rightarrow C_4H_4^+ + C_2H_2 \tag{3c}$$

$$C_6H_6^+ \rightarrow C_3H_3^+ + C_3H_3 \tag{3d}$$

were also detected under these conditions, albeit weakly.[140] When instead the excitation is performed at 253 nm, in resonance with the $6_0^1 1_0^1$ band of the S_1-S_0 transition, the intensity of metastable products of reactions (3c) and

[137] T. Baer, B. P. Tsai, D. Smith, and P. T. Murray, *J. Chem. Phys.*, 1976, **64**, 2460.
[138] H. M. Rosenstock, R. Stockbauer, and A. C. Parr, *J. Chem. Phys.*, 1979, **71**, 3708; *ibid.*, 1980, **73**, 773.
[139] U. Boesl, H. J. Neusser, R. Weinkauf, and E. W. Schlag, *J. Phys. Chem.*, 1982, **86**, 4857.
[140] H. Kühlewind, H. J. Neusser, and E. W. Schlag, *Int. J. Mass Spectrom. Ion Phys.*, 1983, **51**, 255.
[141] H. Kühlewind, H. J. Neusser, and E. W. Schlag, *J. Phys. Chem.*, 1984, **88**, 6104.
[142] H. J. Neusser, H. Kühlewind, U. Boesl, and E. W. Schlag, *Ber. Bunsenges. Phys. Chem.*, 1985, **89**, 276.
[143] H. Kühlewind, A. Kiermeier, and H. J. Neusser, *J. Chem. Phys.*, 1986, **85**, 4427.

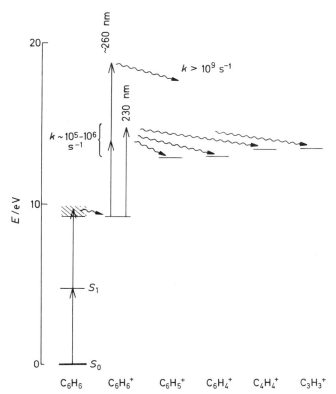

Figure 4 *Energetics for benzene MPIF processes showing neutral, ion, and principal fragment-ion levels*

(3d) increases strongly, indicating an order of magnitude increase in the decomposition rate for these channels at the shorter wavelengths. This of course could be expected for dissociation of a more energetic precursor. A secondary metastable decomposition of these two product-ion species was also identified in these experiments,[140] though this must take place at a net four-photon energy following rapid dissociations, (3c) and (3d).

The dissociations (3a)—(3d) have been of great interest in the chemistry of the benzene ion. Early investigations that used either charge exchange[69] or PEPICO methods[67] to fix the energy of the dissociating system were interpreted to indicate that hydrogen abstraction [reactions (3a) and (3b)] were not competitive with C—C bond cleavage [reactions (3c) and (3d)], with consequent uncertainty as to the identity of the dissociating states. More recent PEPICO data[68] have suggested to the contrary that all four dissociations may compete, although direct evidence was not forthcoming owing to limited mass resolution. The REMPI data for these four dissociations show essentially identical measured appearance rates for each channel at each of the two laser wavelengths and thus strongly suggest that they are competitive fragmentations of a common precursor state.[142]

To investigate this more closely, a modified REMPI ionization scheme was employed.[141,143] Ground-state $C_6H_6^+$ was prepared again by a $(1 + 1)$ scheme but this time using 267 nm photons that excite the neutral through the $^1B_{2u}$ 6_1^0 hot-band. At these wavelengths the two-photon energy exceeds the molecular ionization potential by only 124 meV, and MPI–PES data indicate that, even though this is sufficient to populate some vibrational modes, the ion is produced predominantly in its vibrationless ground state.[144] The internal energy of the prepared ion is therefore more closely defined by this ionization route. Parent benzene cations can then be accelerated into a spatially adjacent region where they are photodissociated by a second tunable laser. This two-colour experiment affords greater control over the dissociation energetics. In the present experiments the four dissociation channels could be investigated over a range of 5.1—5.6 eV internal energy.[141,143] The rotational distribution of the ions was also considered,[143] but because excitation was performed through a rotational bandhead with no obvious structure it was estimated that the resultant ion rotational distribution for these experiments was effectively thermal.

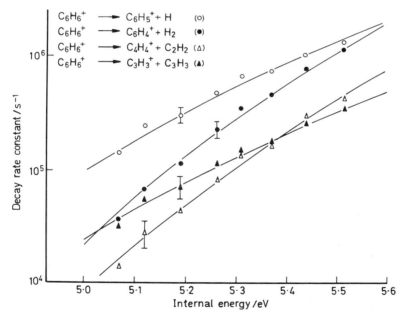

Figure 5 *Unimolecular rate constants, $k(E)$, of the four competing decay channels of benzene cation. The solid lines represent the results of an RRKM calculation fitted to the experimental data*
(Reproduced with permission from *J. Chem. Phys.*, 1986, **85**, 4427)

In this fashion the earlier observation, of a common measured production rate, was confirmed[143] for all four channels. Further, individual $k(E)$ curves

[144] S. R. Long, J. T. Meek, and J. P. Reilly, *J. Chem. Phys.*, 1983, **79**, 3206.

were deduced (Figure 5). This internal-energy-dependent behaviour is well reproduced by RRKM calculations, in which it is assumed that dissociation occurs from the vibrationally excited ground electronic state.[143] The initial population of the ground state must thus be excited to the $^2A_{2u}$ or $^2E_{1u}$ state, then undergo a rapid radiationless transfer back to the ground state prior to fragmentation.

Photofragment Spectroscopy

Spectroscopic investigations of ions have shown a rapid growth in recent years as powerful new techniques have evolved. One aspect has been the development of photofragment spectroscopy,[145] which allows non-fluorescent ions to be examined, provided that the upper state is repulsive or predissociated. Measurement of the fragment intensity as a function of excitation wavelength then permits the level structure to be probed. Particularly when combined with fast beams this method achieves impressively high resolution because of the reduced Doppler width that accrues ('kinematic compression'). However, one drawback is that the ions are typically generated in an electron impact source and emerge in a broad, but undefined, distribution of states. At worst one is then performing spectroscopic experiments where both upper and lower levels are initially unidentified. More generally the situation will be less severe but, for polyatomics especially, spectral congestion arising from the 'hot' source can prove to be a major limitation.

A good example of this is found in a number of investigations of the CH_3I^+ $\tilde{A}$–$\tilde{X}$ laser photodissociation spectrum. Various experiments have demonstrated that vibrational and rotational features are to be observed, yet a detailed assignment has remained elusive.[104,146–150] Chupka, Colson, and co-workers[104,150] have demonstrated the spectral simplification that can be achieved by utilizing REMPI to produce a much narrower distribution of $CH_3I^+(\tilde{X})$. The MPI–PES characterization of the resultant ions has been discussed in the previous section of this chapter. Figure 6 shows the $\tilde{A}^2A_1$–$\tilde{X}^2E_{\frac{3}{2}}$ photodissociation spectra obtained following ion preparation by REMPI via the $v_2 = 0$ and 1 levels of the $6s$ Rydberg state of A symmetry.

According to the MPI–PES data[104] at the longer of the two REMPI wavelengths used here a vibrationless intermediate is ionized to give predominantly a vibrationless, $v_2 = 0$, ground-state ion. This is reflected in the photodissociation spectrum,[104] which can be assigned in terms of $2_0^0 3_0^n$ and

[145] J. T. Moseley, *Adv. Chem. Phys.*, 1985, **60**, 245.

[146] D. C. McGilvery and J. D. Morrison, *J. Chem. Phys.*, 1977, **67**, 368.

[147] S. P. Goss, J. D. Morrison, and D. L. Smith, *J. Chem. Phys.*, 1981, **75**, 757; *ibid.*, 1981, **75**, 1820.

[148] R. G. McLoughlin, J. D. Morrison, D. L. Smith, and A. L. Wahrhaftig, *J. Chem. Phys.*, 1985, **82**, 1237.

[149] P. C. Cosby, unpublished (reported in ref. 145).

[150] W. A. Chupka, S. D. Colson, M. S. Seaver, and A. M. Woodward, *Chem. Phys. Lett.*, 1983, **95**, 171.

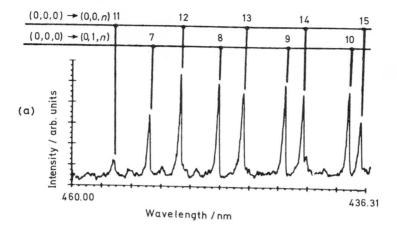

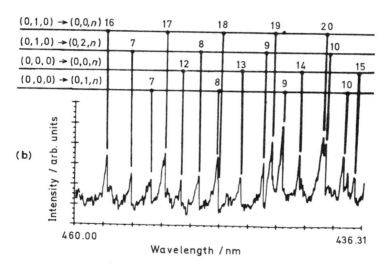

Figure 6 *Photofragment spectrum of* $CH_3I^+(\tilde{X}^2E_{\frac{3}{2}})$ *prepared via* (a) $A(v_2 = 0)$ *and* (b) $A(v_2 = 1)$ *intermediate states. Assignments are given in the form* (v_1, v_2, v_3), *which represent the totally symmetric modes*
(Reproduced with permission from *J. Phys. Chem.* 1986, **90**, 274)

$2_0^1 3_0^n$ progressions (Figure 6a). At the shorter of the two REMPI wavelengths the neutral intermediate has 1 quantum of v_2 excitation and, with the $\Delta v = 0$ ionization predominating, so, too, does the ion. The MPI–PES does, however, indicate some $\Delta v_2 = \pm 1$ ionization. This situation is again reflected in the corresponding photodissociation spectrum (Figure 6b), where two additional series, $2_1^0 3_0^n$ and $2_1^2 3_0^n$, are evident. With these data, together with analogous spectra for photodissociation of the excited $\tilde{X}^2 E_{\frac{1}{2}}$ state ion, a full assignment was attempted.[104]

Little rotational cooling of the ion population was observed in these experiments, other than that afforded by the supersonic expansion of the neutral sample inlet.[150] This is not unexpected for this system because REMPI excitation takes place through rotational bandheads of the neutral Rydberg states, which are observed to be completely featureless.[151] Rotational contour analysis of these bands has been performed and indicates little change in rotational constants between the CH_3I ground and $6s$ Rydberg states.[151,152] Consequently many transitions lie in a given energy interval. A broad rotational distribution is thus formed in the resonant intermediate, which, when ionized, translates into a still broad distribution in the ion.

State-Selected Ion–Molecule Reactions

As a final example of the proven utility of REMPI for state-selective ion generation we present some ion–molecule reaction studies for the following system:[101,153]

$$NH_3^+(v_2) + D_2 \rightarrow NH_3D^+ + D \tag{4a}$$

$$NH_3^+(v_2) + D_2 \rightarrow NH_2D^+ + HD \tag{4b}$$

Reaction (4a) is exothermic by more than 1 eV but has a rate constant that, at $\sim 5 \times 10^{-13}$ cm^3 molecule^{-1} s^{-1}, is considerably less than the Langevin association rate of around 10^{-9} cm^3 molecule^{-1} s^{-1}.[154] Earlier scattering studies using vibrationally unspecified ions indicate a spectator stripping model for this reaction,[155] and a 90 meV activation barrier has been suggested.[154] The exchange reaction (4b) is roughly thermoneutral.

Vibrationally selected ions having a specified number of v_2 quanta were generated in a tandem mass spectrometer by (2 + 1) REMPI, as discussed in the previous section.[100,101,153] As well as allowing the vibrational dependence of reactions (4a) and (4b) to be investigated, the collisional energy could be varied over a range of 0.5—10 eV.

The cross-section for the addition channel, (4a), evidently increases gently with increasing collision energy until ~ 4 eV, whereupon it shows a gradual decline. Correspondingly the exchange channel, (4b), starts to be enhanced above ~ 4 eV.[101] Vibrational excitation ($v_2 = 0$—9), whilst not affecting total product yield, was found to favour (4b) at the expense of (4a). The vibrational dependence of the branching ratios is shown in Figure 7, where it is clear that higher vibrational excitation enhances the exchange channel most effectively. An explanation which is consistent with these observations

[151] D. H. Parker, R. Pandolfi, P. R. Stannard, and M. A. El-Sayed, *Chem. Phys.*, 1980, **45**, 27.

[152] J. A. Dagata, M. A. Scott, and S. P. McGlynn, *J. Chem. Phys.*, 1986, **85**, 5401.

[153] R. J. Morrison, W. E. Conaway, and R. N. Zare, *Chem. Phys. Lett.*, 1985, **113**, 435.

[154] F. C. Fehsenfeld, W. Lindinger, A. L. Schmeltekopf, D. L. Albritton, and E. E. Ferguson, *J. Chem. Phys.*, 1975, **62**, 2001.

[155] G. Eisele, A. Henglein, P. Botschwina, and W. Meyer, *Ber. Bunsenges. Phys. Chem.*, 1974, **78**, 1090.

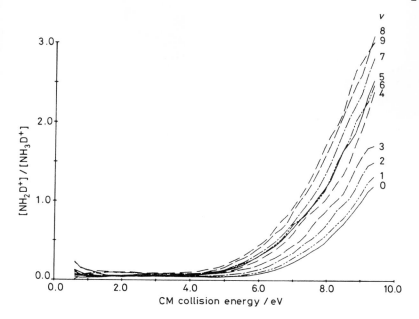

Figure 7 $NH_2D^+:NH_3D^+$ *branching ratio for the reaction* NH_3^+ $(v_2 = 0\text{—}9) + D_2$
showing the dependence on kinetic and v_2 *vibrational energies*
(Reproduced with permission from *J. Chem. Phys.*, 1986, **84**, 5527)

is that vibrational excitation, though not dynamically significant in either reaction channel, increases the yield of NH_2D^+ by promoting dissociation of the initially formed addition product, NH_3D^+. Likewise, excess of kinetic energy will favour this dissociation process.

»